Mathematics as the Science of Patterns

Making the Invisible Visible to Students Through Teaching

Mathematics as the Science of Patterns

Making the Invisible Visible to Students Through Teaching

Edited by

Patrick M. Jenlink
Stephen F. Austin State University

IAP

INFORMATION AGE PUBLISHING, INC.
Charlotte, NC • www.infoagepub.com

Library of Congress Cataloging-in-Publication Data

CIP record for this book is available from the Library of Congress
http://www.loc.gov

ISBNs: 978-1-64802-744-4 (Paperback)

978-1-64802-745-1 (Hardcover)

978-1-64802-746-8 (ebook)

Printed and bound by CPI Group (UK) Ltd, Croydon, CR0 4YY

Dedication

To mathematics researchers and educators and the future generations of mathematics teachers and students who will require an understanding of and appreciation for *mathematics as the science of patterns* in understanding and applying mathematical knowledge in a changing world.

CONTENTS

PREFACE

Mathematics has historically been called a *science of patterns*. We think of mathematics as having structure, and that structure enables us to solve problems. The structure is built around looking for and manipulating patterns. Many argue that patterns are the cornerstone of mathematics. They are the foundation that the whole of the subject is built on. From the earliest tally systems to the development of differential calculus to modern mathematics patterns were and are the genesis, the motivation, and the foundation of mathematical knowledge. As such mathematics is often referred to as the *science of patterns*.

We are in a time of great interest in mathematics, in particular K–12 mathematics education. Student performance, curriculum, and teacher education are the subjects of much scrutiny and debate. Studies of the mathematical knowledge of prospective and practicing teachers, both in the United States and globally, suggest ways to improve their mathematical educations. Importantly, understanding patterns in mathematics, on the part of the teacher and student learner, plays a critical role in learning to teach for teachers, and equally important, learning mathematics for students.

The value of patterns to the teaching and learning of mathematics is well understood, both in terms of research and application. When we involve or appeal to pattern in teaching mathematics, it is usually because we are trying to help students to extract greater meaning, or enjoyment, or both, from the experience of learning environments within which they are

Mathematics as the Science of Patterns:
Making the Invisible Visible to Students Through Teaching, pp. ix–xii

occupied, and perhaps also to facilitate remembering. As a general skill it is thought that the ability to discern a pattern is a precursor to the ability to generalize and abstract, a skill essential in the early years of learning and beyond.

The formal treatment of patterns in primary years is focused initially on repeating patterns. A repeating pattern is a pattern that has a discernable unit of repeat. In many ways this is a formalization of the patterns that students have begun to experience in the cyclic nature of the days of the week, months of the year, and hours of the day. Research has demonstrated that making patterns that are often cognitively invisible to students is critical for students being able to see and learn patterns necessary to solve mathematical problems. This is particularly important at early ages in addition, subtraction, multiplication, division, algebra, and beyond into adolescent and college students experiencing more complex mathematics.

Pattern recognition in mathematics (the detection of regularity) is one of the six basic processing skills (the others are the abilities to: particularize, interpret, describe, make logical inferences, and identify a set of axioms). Patterning has been claimed as an aesthetic mode of cognition by which we constantly and successfully make sense of our environment.

Equally important, the role of patterns as pedagogical tools cannot be overlooked. It is often through the use of patterns that a teacher is able to unlock the truth within mathematical theorems and proofs. It is also through the use of patterns that students are able to explore new ideas. The role of work with repeating patterns is a useful basis for teaching about other matters, with the pattern-making acting as a concrete and familiar experience, which can be meaningfully referred to in talking about new ideas.

Research indicates that the larger problem in teaching mathematics does not lie primarily with students; rather it is with the teachers themselves. In order to make changes for students there first needs to be a process of change for teachers. Understanding the place of patterns in learning mathematics is a predicate to understanding how to teach mathematics and how to use pedagogical reasoning necessary in teaching mathematics. Importantly, the lack of distinction created by the pedagogical use of patterns is not immediately problematic to the student or the teacher. The deep-seated cognitive patterns that both teachers and students bring to the classroom require change.

For preservice teachers this would best be facilitated in the treatment of the topic of patterns in both their mathematics preparation (Foundations of Mathematics courses) and their mathematics teaching preparation (Methods of Teaching Mathematics courses). As an example, in the foundation course, dissatisfaction with their current way of thinking can be facilitated with the use of pattern-based problem-solving activities. In the methods course teacher educators would introduce preservice teachers to mathematical problems, situated in a dynamic experiential context of discourse focused on disclosing and then identifying old cognitive patters to be replaced by a new understanding of the patterns in mathematics.

Teacher educators that are focused on mathematics thinking and understanding can provide valuable insights and information about what takes place in school classrooms, including common mathematical misunderstandings of practicing teachers and how to build on these to promote real understanding. This strengthens the opportunity for new teachers to be successful in teaching mathematics. As concepts like fractions and decimals enter the curriculum, teaching mathematics well requires subject matter expertise that nonspecialists cannot be expected to master. Research in mathematics, particularly research focused on patterns in learning, cognition, and reasoning as well as nuances in teaching mathematics are important to preparing future mathematics classroom teachers.

Mathematics as the Science of Patterns: Making the Invisible Visible to Students through Teaching introduces the reader to a collection of thoughtful, research-based works by authors that represent current thinking about mathematics, mathematics education, and the preparation of mathematics teachers. Each chapter focuses on mathematics teaching and the preparation of teachers who will enter classrooms to instruct the next generation of students in mathematics.

Chapter 1 opens the book with a focus on mathematics as the science of patterns and the importance of patterns in mathematical problem solving, providing the reader with an introduction. The authors of Chapter 2 revisit the work of Pólya and the development and implementation of problem solving in mathematics. In Chapter 3, the authors present an argument for core pedagogical content knowledge in mathematics teacher preparation. The authors of Chapter 4 focus on preservice teachers' patterns of conception as related to understanding number and operation. In Chapter 5 the authors examine the role of visual representation in exploring proportional reasoning, denoting the importance of each in helping learners make their thinking visible. The authors of Chapter 6 examine patterns and relationships, and the importance of each in assisting students' learning and development in mathematical understanding. The authors of Chapter 7 examine the use of worked examples as a scalable practice, with emphasis on the importance of worked examples in teaching fraction magnitude and computation is discussed. In Chapter 8, the authors expand on the zone of proximal development to investigate the potential of Zankov's Lesson in terms of students analyzing numerical equalities. The authors of Chapter 9 focus on high leverage mathematical practices in elementary preservice teacher preparation, drawing into specific relief the APEX cycle to develop deep thinking. In Chapter 10, the author focuses on number talks and the engagement of students in mathematical reasoning, which provides opportunities for students to be sensemakers of mathematics. Chapter 11 presents an epilogue, focusing on the importance of recognizing the special nature of mathematics knowledge for teaching.

ACKNOWLEDGMENTS

The initial idea for this book began as a conversation among colleagues about the fundamental idea of mathematics as *the science of patterns*. In particular, we focused on the importance of "patterns" as related to understanding and applying "mathematical knowledge for teaching and learning." Equally influential was a focus on *the science of patterns* in relationship to teachers' mathematical thinking, mathematical content knowledge and pedagogical content knowledge. The concern we fostered was the complex nature of cognitive reasoning and mathematical patterns that pervade mathematical understanding in relation to teaching and learning. Specifically, we discussed preparation of the future generations of mathematics teachers and the need for their being prepared to teach students of mathematics. This, we believed, required a focus on research-based understandings of mathematics teaching and learning.

In our conversations, we asked what worked and didn't work and why with respect to the deep knowledge of patterns on mathematics and the teaching and learning of mathematics. We recognized that using a subject-specific framework for researching teachers' knowledge could give researchers, teacher educators, and teacher practitioners a window into the ways in which teacher knowledge influences students' learning mathematics. With gratitude for my colleagues I acknowledge contribution.

Acknowledgement and thanks goes the contributing authors whose research offers insight and thoughtful considerations for understanding the need to examine the inherent patterns in mathematical problems and

Mathematics as the Science of Patterns:
Making the Invisible Visible to Students Through Teaching, pp. xiii–xiv

the knowledge of mathematics required for teaching students how to solve those mathematical problems.

The authors bring the welcomed perspective of theorists and researchers as well as their own field-based research to the discourses presented in the book. Without the contributing authors, this book would not have been possible. The authors of the chapters examining mathematics, presenting insight into the patterns of mathematics, and important role of teachers bring their considerable experience to bear on interpreting the complexity, challenges, and problems associated with preparing each new generation of mathematics teachers.

Gratitude is extended to the external reviewers who took time out their busy schedules to review and provide comments and suggestions on the chapters. The reviewers brought a deep-seated understanding of patterns in mathematical reasoning and problem solving. Acknowledging the value of the chapters and offering constructive feedback was invaluable, as was the affirmation by reviewers of both the need for and importance of a book committed to examining the intricacies of *mathematics as a science of patterns* in teaching and learning mathematics.

Likewise, gratitude is extended to George Johnson and his colleagues at Information Age Publishing (IAP) for their vision in seeing the value of a book on *mathematics as the science of patterns* that draws into specific relief the need to advance an understanding of how the invisible patterns involved in problem solving are made visible through teaching of mathematics.

As well, thanks the production staff at IAP is in order for their ever-vigilant efforts to move the book through to completion. Working with a quality publisher and the folks that do the work to translate a manuscript into a completed book is a rewarding experience.

Finally, gratitude goes to Stephen F. Austin State University for supporting this project and enabling the realization of a work that will shape educator preparation for years to come.

CHAPTER 1

MATHEMATICS—THE SCIENCE OF PATTERNS

The Importance of Patterns in the Teaching and Learning of Mathematics

Patrick M. Jenlink

Introduction

Preparing mathematics teachers and teacher educators is a complex, "interdisciplinary enterprise requiring knowledge of teaching and learning as well as knowledge of mathematics" (Conference Board of the Mathematical Sciences [CBMS], 2012, p. xi). The level of mathematics knowledge for teaching teachers[1] required to prepare teachers to enter classrooms in schools and teach mathematics has increased exponentially, not withstanding the level of mathematics knowledge required to successfully teach mathematics to diverse student populations with varied mathematical experiences in learning and seeking to understanding mathematics.

Given that mathematics education places emphasis on a high level of mathematical knowledge, inclusive of the preparation of mathematics teacher educators, mathematics teachers, and the teaching of mathematics situated in the context of schools and universities, one need only reflect on

Mathematics as the Science of Patterns:
Making the Invisible Visible to Students Through Teaching, pp. 1–16

the last five decades forward to recognize that an emphasis on mathematics as a study of numbers and number theory was a driving philosophical force on how mathematics was defined and mathematics curricula and teaching were guided. Importantly, the past three decades or so have witnessed a transition to include an increasing emphasis on mathematics as the study and practice of patterns in problem solving.[2] Halmos (1980) writes that the work of the mathematician "is to solve problems, and that, therefore, what mathematics really consists of is problems and solutions" (p. 519). Therein, mathematics education, the teaching of mathematics that focuses on solving problems, draws forward the importance of understanding patterns in mathematics problem solving.

Along a similar trajectory, an increasing number of mathematicians, mathematics researchers, and mathematics educators have recognized the special nature of the mathematical knowledge needed for K–12 teaching and its implications for the mathematical preparation of teachers (CBMS, 2001).[3] Simply stated, mathematics makes sense of day-to-day activities that require an understanding of patterns. Mathematics enables us "to perceive patterns, to comprehend data, and to reason carefully" (National Research Council [NCR], 1989, p. 43). Revealing hidden patterns through mathematics helps us understand and negotiate the world around us. Increasingly, over the past two decades, mathematics has become expanded beyond arithmetic and geometry, rather, in today's world mathematics is understood as a diverse discipline that deals with data, measurements, and observations from science; with inference, deduction, and proof; and with mathematical models of natural phenomena, of human behavior, and of social systems (NRC, 1989, 1990, 2001).

Mathematics is increasingly perceived, understood, and practiced as a *science of patterns* (Devlin, 1994; Oliveri, 1997; NRC, 1989; Resnik, 1981, 1982; Schoenfeld, 1992; Steen, 1988; Wittmann, 2005). Patterns are evident in every aspect of mathematics. The mathematician, Devlin (1997) examines abstract patterns including "numerical patterns, patterns of shape, patterns of motion, patterns of behavior, and so on" (p. 3). Devlin further explains that

> mathematical activity is concerned with finding new patterns in the world, analyzing those patterns, formulating axioms to describe them and facilitate further study, looking for the appearance in a new domain of patterns observed somewhere else, and applying mathematical theories. (p. 52)

Most importantly, mathematics as a *science of patterns* is a science of abstract objects that relies on plausible reasoning (Pólya, 1954) as its standard of truth, yet employs observation, simulation, and even experimentation in problem solving as means of discovering truth. To teach mathematics to students an understanding of the special role of mathematics in education that is a consequence of its universal applicability.

Importantly, what mathematics as a *science of patterns* means, for the mathematician and mathematics researcher, is that to know mathematics is to investigate and express relationships among patterns: to be able to discern patterns in complex and obscure contexts; to understand and transform relations among patterns; to classify, encode, and describe patterns; to read and write in the language of patterns; and to employ knowledge of patterns for various practical purposes. Likewise, for the mathematics teacher, mathematics as a *science of patterns* requires having the mathematical knowledge[4] and mathematical reasoning acumen to teach students about patterns, beginning at an early age and continuing to incorporate pattern reasoning and problem solving through high school and into college.

Focusing on the relationship of mathematics as a *science of patterns* and the preparation of mathematics teachers, the teacher educator and the preservice teacher of mathematics need to understand that "mathematics is a living subject which seeks to understand patterns that permeate both the world around us and the mind within us" (NRC, 1989, p. 84). This requires a change in mindset, the habits of mind one brings to teaching, that sees the traditional language of mathematics as "based on rules that must be learned" (NRC, 1989, p. 84) and move beyond rules to be able to see mathematics as a *science of patterns*, not as a canon, and to recognize that mathematics is really about patterns and not entirely about numbers. This shift in thinking and understanding requires an effort for teachers to focus on teaching mathematics wherein the teacher guides the students in seeking solutions, not just memorizing procedures; exploring patterns, not just memorizing formulas; and formulating conjectures, not just doing exercises (NRC, 1989, p. 84).

Mathematical Patterns in Teaching Mathematics

Given the understanding put forward for mathematics as a *science of patterns*, research in mathematics, mathematics education, teacher education, and teaching mathematics practice in recent years has focused increasingly on patterns, patterning, and pattern tasks, drawing attention to the importance of understanding patterns in mathematics teaching and learning (Amit & Neria, 2008a, 2008b; Barbosa et al., 2012; Fujita & Yamamoto, 2011; Gilmore & Papadatou-Pastou, 2009; Lannin, 2005; Maglipong & Bongolto, 2017; Markworth, 2016; Michael et al., 2006; Mulligan & Mitchelmore, 2009; Noss et al., 1997; Orton et al., 1999; Tsamir et al., 2017; Vanluydt et al., 2021; Vogel, 2005; Warren, 2005; Wittmann, 2005; Zazkis & Liljedahl, 2002; Zopf, 2010).

Searching for patterns in different contexts, using and understanding symbols and variables that represent patterns, and generalizing are significant components of the mathematics curriculum in many countries and the importance of an emphasis on pattern reasoning in mathematics teacher preparation (Barbosa et al., 2008). Specifically, the importance of developing teacher mathematical knowledge focuses on abilities related to searching and exploring numeric and geometric patterns, as well as solving problems, looking for regularities, conjecturing and generalizing. The search for patterns is seen as important to teaching students in early pre-algebra mathematics as a crucial step for establishing generalization, which is the essence of mathematics (Barbosa et al., 2008; Noss et al., 1997; Orton & Orton, 1999; Orton et al., 1999; Warren, 2005; Zazkis & Liljedahl, 2002).

Maglipong and Bongolto (2017) argue that "the use of mathematical patterns is an effective method of teaching in improving students' achievement and conceptual understanding" (p. 2012). Number patterns, the relationship between variables and generalization, are considered important components of mathematics teaching, and student learning and functional thinking (Warren et al., 2006). Patterns can suggest numerical, visual, and mixed approaches (Stacey, 1989; Stylianous & Silver, 2004). Research recommends that mathematical patterns may be used by mathematics teachers in their classes to help improve achievement and develop conceptual understanding even to preschool, elementary and secondary levels. Stylianides and Silver (2009), in their research, note that pattern identification is integral to learning mathematics and plays a critical role in students' mathematical reasoning and proving in problem solving (Vale & Cabrita, 2008).

Prediger (2010), addressing the question, "What kind of mathematical knowledge do prospective teachers need for teaching and for understanding student thinking?," specifically focused on mathematics knowledge for teaching by investigating the case of understanding students' perspectives on equations and on meanings of equations and the equal sign. Considering that "meaning"[5] is an important part of understanding the kind of mathematics knowledge for teaching required, Bass and Ball (2004), offer the insight

> that, whereas the work of research mathematicians might be described in terms of "compressing" information into increasingly concise and powerful formulations, the work of teachers is more often just the opposite: teachers must be adept at prying apart concepts, making sense of the analogies, metaphors, images, and logical constructs that give shape to a mathematical construct. (Davis & Simmt 2006, p. 300)

Prediger's (2010) study resulted in "the analysis of student thinking as an opportunity to construct didactically-sensitive mathematical knowledge-for-teaching for pre-service middle-school mathematics teachers" (p. 73). Importantly, this study demonstrated that diagnostic competence comprises didactically sensitive mathematical knowledge for teachers.

Research concerning methods of generalization has focused on generalizing patterns, which vary in types of patterns, numerical, pictorial or repeating patterns, and differ in population, from children to preservice school teachers (e.g., Amit & Neria, 2008b; Rivera & Becker, 2005; Stacy, 1989; Zazkis & Lijendak, 2002). Amit and Neria (2008b) in their research of the value of pattern problems in learning mathematics found that pattern problems designed to engage the mathematics learner in pattern problems "have been found to be efficient in developing and revealing the ability to generalize" (p. 49). The results of their study demonstrated the importance of exposing mathematics students "to non-linear patterns, since they increase the challenge of generalization, provide novel mathematical experiences, and have the potential to enhance mathematical development" (p. 49).

Teaching mathematics through approaches involving patterning and problem solving[6] activities requires thoughtful and well-executed design to accommodate students' learning of mathematics. Durand-Guerrier et al. (2012) note that "engaging students in situations which make them aware of the constructive character of mathematical activities, especially those involving conjecture and proof, possess complex challenges" (p. 364). Jahnke (2007) reflects that student learning, in mathematics classroom exercises, through observation and conjecturing followed by explanations, is invaluable for students learning of mathematics, importantly because it situates students actively in the process of knowledge creation and discovery new mathematical relations.

Problem Solving—A Focus on Pattern and Patterning

Attention in recent years has been directed toward fostering deep mathematical knowledge in preservice teachers by engaging them in learning mathematics through mathematics problem solving[7] (Lester, 2013; Masingila, Lester, & Raymond, 2011a; Masingila et al., 2011; Masingila, Olanoff, & Kimani, 2018; Stein et al., 2003). Learning through problem solving is challenging and requires both the mathematics teacher educator and the mathematics preservice teacher to renegotiate roles in teaching and learning mathematics. Important consideration is needed in teaching mathematics through problem solving. Specifically, understanding the complexity of how one learns to teach and what is required, developed, and used in supporting preservice teachers learning mathematical knowledge for teaching is complex.

Considering mathematical knowledge for teaching, it is important to recognize that all mathematics in todays' mathematical world is based on pattern and structure.[8] Teaching the preservice teacher to enter a classroom to teach mathematics requires, at a minimum, a high level of mathematical reasoning and explaining, modeling, being able to see structure, and generalizing. With that in mind, the mathematics community has promoted engaging preservice teachers in problem solving to support the development of deep mathematical understanding.

As Warren (2005) asserts, "the power of mathematics lies in relations and transformations that give rise to patterns and generalizations. Abstracting patterns is the basis of structural knowledge, the goal of mathematics learning" (p. 305). Teaching mathematics should focus on developing "fundamental skills in generalising, and expressing and systematically justifying generalisations" (p. 305).[9] There are many indications that an understanding of pattern and structure is important in early mathematics learning (Mulligan & Mitchelmore, 2009). Research focused on visualisation skills, which frequently involve recognition of pattern and structure, are positively correlated with mathematical achievement (Arcavi, 2003). Relatedly, research on imagery and arithmetic has indicated that students who recognise the structure of mathematical processes and representations acquire deep conceptual understanding (Gray et al., 2000; Pitt-Pantazi et al., 2004). Whereas students with lower numerical achievement reported descriptive and idiosyncratic images as they focused on nonmathematical aspects and surface characteristics of visual cues

Noss et al. (1997) researched the construction of mathematical meanings related to connecting the visual with the symbolic. Specifically, their research focused on "mathematical meanings derive from connections" (p. 203). Their premise was that learning mathematics begins with the concrete, and "ascends" (the metaphor is pertinent) to the abstract. Utilizing a "matchstick" mathematics problem wherein students were expected to identify relationships within numerical patterns of matchstick placement patterns derived from spatial or practical situations as the progression of matchsticks proceeded. The intent is for the students to express in natural language the patterns and ultimately formulate a symbolic generalization. The results reported "intra-mathematical connections which link new mathematical knowledge with old, shaping it into a part of the mathematical system; and extra-mathematical meaning derived from contexts and settings which include—though not uniquely—the experiential world" (Noss et al., 1997, p. 203).

Markworth (2016), in a study on repeating patterns,[10] notes that the "mathematical connections that can be made between primary work with repeating patterns and content that is developed in later years" (p. 28) is important in the learning of mathematics. Repeating patterns help students see their world mathematically. And, regardless of the solution strategy or the mathematical problem, the problem and solution "cannot be accessed unless the problem solver is able to recognize, duplicate, and extend the repeating pattern" (p. 28). Problem solving tasks challenge students and set high expectations for the development and use of mathematical thinking skills that involve communication, conjecture, generalization, argumentation, and proof; searching for and identifying patterns is an important problem-solving strategy

Lüken (2020) has noted, in research on patterning as a mathematical activity, that in both educational and research settings on repeating patterns, the focus seems to be put more on the product (i.e., the correct solution) than on the process (i.e., the strategies children employ in solving a patterning task). The research explores different patterning strategies and strategy use in early childhood and reflects on selected patterning tasks from a strategy point of view. Results of the study demonstrated that young children use a variety of different patterning strategies, with the older children using more advanced strategies more often. Lüken (2020) notes, "when considering the mathematical structure of repeating patterns, it becomes apparent that most young children approach patterning activities without referring to or using the unit of repeat" (p. 79).

Pattern Tasks and Problem Solving

The perspective of problem solving, as Mamona-Downs and Downs (2013) explain, "has a relatively compact core of ideas, mainly centered on heuristics, meta-cognition including executive control, accessing and applying the knowledge base, and identifying patterns of modes of thinking as students' work progresses" (p. 138). Following the pioneering work of Pólya (e.g., Pólya, 1957) and later by Schoenfeld (1992), problem solving as a domain of mathematical activity "concerns the student's engagement on any mathematical task that is not judged procedural or the student does not have an initial overall idea how to proceed in solving the task (Mamona-Downs & Downs, p. 138). A problem-solving approach adopted in teaching (either for mathematics teacher preparation or teaching mathematics in school classrooms) has a number of issues about problem design and solution and/or proof assessment. Relatedly, there are questions that guide the design: Mamona-Downs and Downs (2013) ask, "What constitutes a 'good' problem? For this question, one could simply say that any problem for which there is not immediately an obvious line of attack is suitable. But do other factors come in?" (p. 140). Responses included "A sense of the attractiveness of the solution is one, a sense of achievement the solver experiences is another" (p. 149).[11]

Importantly, in designing a pattern task, the mathematics teacher educator or the mathematics teacher is responsible for selecting and facilitating pattern tasks that are appropriate for the mathematical problem solving that he or she wants the preservice students to engage in, tasks designed to learn mathematics (Masingila et al., 2018). Likewise, when the student of teaching mathematics matriculates into being a teacher of mathematics in the school classroom to teach mathematics, then she or he is responsible for selecting and facilitating pattern tasks for her or his students. At both levels of teaching and learning mathematics, the university teacher preparation program and the teacher of mathematics in the educational environment in the school, involves choosing high-level tasks and implementing the tasks that are appropriate for the level of learner or student "in ways that maintain the cognitive demand of the tasks" (Henningsen & Stein, 1997).

High-level tasks engage students in cognitive reasoning, making connections among mathematical concepts and working through mathematical problems, making and testing conjectures, justifying solutions, solving problems for which a procedure is not known, providing solutions and proofs, while low-level tasks involve memorization and procedures that are not connected to conceptual ideas and less focused on identifying and reasoning of mathematical patterns necessary for solving the task (Henningsen & Stein, 1997). Facilitating students' engagement with high-level tasks and maintaining the cognitive demand of tasks requires both a deep understanding of mathematics knowledge as well as pedagogical skills from the mathematics teacher (again, either the mathematics teacher educator or the mathematics teacher entering the classroom). Reflection on their work as mathematics teacher educators and/or as mathematics teachers, as Chapman (2008) explains, includes "examining, framing, and attempting to solve the dilemmas of classroom practice; and being aware of and questioning the assumptions and values he or she brings to teaching" (p. 121). Some of these skills include scaffolding students' thinking, along with creating an environment where student explanations are expected.

Lester (2013), in reflecting on mathematical pattern tasks and problem-solving design, identified five types of features that serve to describe problem tasks: syntax, content, context, structure, and process. The syntax feature refers "to the arrangement of and relationships among words and symbols in a task" while the content feature deals with "the mathematical meanings in the problem" (p. 265). Lester clarifies "two important categories of content features are the mathematical content area (e.g., geometry, probability) and linguistic content features (e.g., terms having special mathematical meanings such as 'less than,' 'function,' 'squared')." Lester distinguishes content features as "the non-mathematical meanings in the task statement. Furthermore, context features describe the problem embodiment (representation), verbal setting, and the format of the information given in the problem statement" (pp. 265–266).

Concerning the structure feature, Lester (2013) explains this feature can "be described as the logical-mathematical properties of a task. Structure features are determined by the particular representation that is chosen for a problem" (p. 266). As an example of the structure feature, "one student may choose to represent a task in terms of a system of equations, while another student may represent the same problem in terms of some sort of guessing process" (p. 266). Finally, Lester explains the process feature as representing "something of an interaction between task and student. That is, although problem-solving processes (e.g., heuristic reasoning) typically are considered characteristics of the student, it is reasonable to suggest that a problem may lend itself to solution via particular processes" (p. 266). As Lester notes, "a consideration of task process features can be very informative to the researcher in selecting tasks for both instruction and assessment" (p. 266).

Conclusion

Teaching educators teaching mathematics to prospective teachers, as well as the preservice teacher transitioning to a mathematics teaching position in a school, is both a complex and important task concerning the role responsibility of teaching mathematics in todays' educational environments. The recognition that the mathematics of 50 years ago has changed in terms of how it is viewed by mathematicians, mathematics researchers, mathematics teacher educators, and mathematics teachers as a *science of patterns*. This change places new demands, in terms of mathematical knowledge as well as instructional considerations, on everyone committed to educating the future mathematics generations, perhaps more so on those preparing mathematics teachers in the university setting and on those entering school classrooms to teach mathematics.

Problem solving and mathematical tasks are increasingly more important to the education of mathematics teacher educators and the mathematics teachers focused on teaching mathematics to every diverse population of mathematics students whose knowledge and experience in terms of learning mathematics is of primary concern. This is equally true for all grade levels in K–12 schools as it is for higher education. The value of patterns in problem solving to the teaching and learning of mathematics is well researched and understood. Mathematics as a *science of patterns* brings new opportunities to the work of teaching mathematics, and preparing the next generation of mathematicians, mathematics researchers, mathematics teacher educators, and mathematics teachers for the future.

NOTES

1. Zopf (2010) defined mathematics knowledge for teaching teachers as, "the mathematical knowledge used by mathematics teacher educators in the work of teaching mathematics to teachers" (p. 11), and argues that, the major purpose of the work of [mathematics] teacher education is beginning with people who already know some mathematics and developing that knowledge into mathematical knowledge for teaching" (p. 165).
2. General problem-solving strategies are not only learnable and teachable but are an essential element of mathematical knowledge. Perhaps the best known position on this view was provided by Pólya (1957) when he discussed a range of general problem-solving strategies (a heuristic), such as encouraging mathematics students to think of a related problem and then solve the current problem by analogy or to think of a simpler problem and then extrapolate to the current problem.
3. Mathematicians are particularly concerned with teachers' knowledge and understanding of mathematics. Recognition and support of the role and responsibility of mathematicians in this area was highlighted by the publication by the CBMS (an umbrella organization for the U.S. professional societies in mathematics) of The Mathematical Education of Teachers (CBMS, 2001), a report that stimulated widespread and continuing efforts to strengthen the mathematical preparation of teachers.
4. Deborah Ball and her research groups at the University of Michigan have provided a research-based emerging theory of what has been named mathematical knowledge for teaching (MKT) focused on answering questions such as: What do teachers need to know, in what ways, and for what purposes? And how can they gain such knowledge? Their research has also been concerned with the knowledge required of teachers to teach mathematics, which must simultaneously respect the integrity of the mathematical ideas and also attend closely to the mathematical development of their students (Ball et al., 2008; Bass & Ball, 2004; Hill et al., 2008; Hill & Charalambous, 2012; Hill et al., 2005; Hill et al., 2004).
5. Prediger denotes that for the mathematical content chosen for the study, the equal sign or equality, there has been a long tradition of reflecting and researching on meaning, which was described as a process of broadening the sources of meaning (see Kieran, 2006; Kilpatrick et al., 2005 for further discussion).
6. Lester (2013) notes that for problem solving, either the design of the problem or engaging in solving the problem requires "ample relevant experience in learning how to solve problems, strong content knowledge, proficiency in using a variety of representations and a solid grasp of how to recognize and construct patterns of inference" (p. 248). Consequently, the teacher must embody these requirements and be able to facilitate the mathematics student in ways of understanding the problem as well as engaging in solving the problem.
7. Lesh and Zawojewski (2007) define problem solving as: "A task, or goal-directed activity, becomes a problem (or problematic) when the 'problem

solver' (which may be a collaborating group of specialists) needs to develop a more productive way of thinking about the given situation" (p. 782). Lesh and Zawojewski argued that problem solving should not be separated "from concept development and the ways that these concepts are used in ... situations beyond school" (p. 782).

8. Harel (2008) explains "structure" in that mathematics is defined as a discipline consisting of these two sets of knowledge. Specifically: "Mathematics is a union of two sets: The first set is a collection, or structure, of structures consisting of particular axioms, definitions, theorems, proofs, problems, and solutions. This subset consists of all the institutionalized ways of understanding in mathematics throughout history. The second set consists of all the ways of thinking that are characteristics of the mental acts whose products comprise the first set." The main pedagogical implication of this definition is that mathematics curricula at all grade levels, including curricula for teachers, should be thought of in terms of the constituent elements of mathematics (p. 490).
9. Such experiences give rise to understandings that are independent of the numbers or objects being operated on (e.g., a + b = b + a regardless of whether a and b are whole numbers, decimals, or variables). Ohlsson (1993) names such understanding abstract schema and argues they are more likely to promote transfer to other mathematical notions than a schema based on particular numbers or content. (Warren, 2005, p. 305).
10. Markworth (2016) defines a *repeating pattern* as "a cyclical repetition of an identifiable core. Children in the primary grades usually begin pattern work with fairly simple patterns, such as AB, ABC, or ABB patterns" (p. 25).
11. A measure of a "good" problem is how far a solution, or an attempt to achieve a solution, would inspire the solver to form related problems.

REFERENCES

Arcavi, A. (2003). The role of visual representations in the learning of mathematics. *Educational Studies in Mathematics*, *52*, 215–241. https://doi.org/10.1023/A:1024312321077

Amit, M.. & Neria, D. (2008a). "Rising to the challenge": Using generalization in pattern problems to unearth the algebraic skills of talented pre-algebra students. *ZDM*, *40*(1), 111–129. https://doi.org/10.1007/s11858-007-0069-5

Amit, M., & Neria, D. (2008b). Methods for the generalizations of non-linear patterns used by talented pre-algebra students. In O. Figueras, J. L., Cortina, S., Alatorre, T. Rojano, & A. Sepúlveda (Eds.), *Proceedings of the Joint Meeting of PME 32 and PME-NA XXX* (Vol. 2. pp. 49–56). Cinvestav-UMSN.

Ball, D. L., Thames, M. H., & Phelps, G. (2008). Content knowledge for teaching: What makes it special? *Journal of Teacher Education*, *59*, 389–407. https://doi.org/10.1177/0022487108324554

Barbosa, A., Vale, I., & Palhares, P. (2012). Pattern tasks: Thinking processes used by 6th grade students. *Revista Latinoamericana de Investigación en Matemática Educativa*, *15*(3), 273–293.

Barbosa, A., Vale, I., Palhares, P. (2008). The influence of visual strategies in generalization: a study with 6th grade students solving a pattern task. In K. Kumpulainen & A. Toom (Eds.), *ETEN 18 The Proceedings of the18th Conference of the European Teacher Education Network hosted by Liverpool Hope University* (pp. 71–77). Helsinki, Finland: University of Helsinki.

Bass, H., & Ball, D. L. (2004). A practice-based theory of mathematical knowledge for teaching: The case of mathematical reasoning. In W. Jianpan & X. Binyan (Eds.), *Trends and challenges in mathematics education* (pp. 107–123). East China Normal University Press.

Chapman, O. (2008). Mathematics teacher educators' learning from research on their instructional practices: A cognitive perspective. In B. Jaworski & T. Wood (Eds.), *The international handbook of mathematics teacher education: The mathematics teacher educator as a developing professional* (Vol. 4, pp. 115–134). Sense.

Conference Board of the Mathematical Sciences (CBMS). (2001). *The mathematical education of teachers*. American Mathematical Society.

Conference Board of the Mathematical Sciences (CBMS). (2012). *The mathematical education of teachers II*. American Mathematical Society and Mathematical Association of America.

Davis, B., & Simmt, E. (2006). Mathematics-for-teaching: An ongoing investigation of the mathematics that teachers (need to) know. *Educational Studies in Mathematics, 61*, 293–319. https://doi.org/10.1007/s10649-006-2372-4

Devlin, K. (1994). *Mathematics: The science of patterns: The search for order, life, mind and order in the universe*. Scientific American Library.

Durand-Guurrier V., Boero P., Douek N., Epp S. S., & Tanguay D. (2012). Argumentation and proof in the mathematics classroom. In G. Hanna & M. DeVilliers (Eds.), *Proof and proving in mathematics education. The 19th ICMI study* (pp. 349–368), Springer.

Fujita, T., & Yamamoto, S. (2011). The development of children's understanding of mathematical patterns through mathematical activities. *Research in Mathematics Education, 13*(3), 249–267. https://doi.org/10.1080/14794802.2011.624730

Gilmore, C. K., & Papadatou-Pastou, M. (2009). Patterns of individual differences in conceptual understanding and arithmetical skill: A meta-analysis. *Mathematical Thinking and Learning, 11*, 25–40. https://doi.org/10.1080/10986060802583923

Gray, E., Pitta, D., & Tall, D. O. (2000). Objects, actions, and images: A perspective on early number development. *Journal of Mathematical Behavior, 18*, 401–413. https://doi.org/10.1016/S0732-3123(00)00025-0

Halmos, P. R. (1980). The heart of mathematics. *American Mathematical Monthly, 87*, 519–524. https://doi.org/10.1080/00029890.1980.11995081

Harel, G. (2008). DNR perspective on mathematics curriculum and instruction, Part I: focus on proving. *ZDM Mathematics Education, 40*, 487–500. https://doi.org/10.1007/s11858-008-0104-1

Henningsen, M., & Stein, M. (1997). Mathematical tasks and student cognition: Classroom-based factors that support and inhibit high-level mathematical thinking and reasoning. *Journal of Research in Mathematics Education, 28*, 524–549. https://doi.org/10.2307/749690

Hill, H. C., Blunk, M. L., Charalambous, C. Y., Lewis, J. M., Phelps, G. C., Sleep, L., & Ball, D. L. (2008). Mathematical knowledge for teaching and the mathematical quality of Instruction: An exploratory study. *Cognition and Instruction, 26*(4), 430–511. https://doi.org/10.1080/07370000802177235

Hill, H. C., & Charalambous, C. Y. (2012). Teacher knowledge, curriculum materials, and quality of instruction: Lessons learned and open issues. *Journal of Curriculum Studies, 44*(4), 559–576. https://doi.org/10.1080/00220272.2012.716978

Hill, H. C., Rowan, B., & Ball, D. L. (2005). Effects of teachers' mathematical knowledge for teaching on student achievement. *American Educational Research Journal, 42*(2), 371–406. https://doi.org/10.3102/00028312042002371

Hill, H., Schilling, S., & Ball, D. (2004). Developing measures of teachers' mathematics knowledge for teaching. *The Elementary School Journal, 105*(1), 11–30. https://doi.org/10.1086/428763

Jahnke, H. N. (2007). Proofs and hypotheses. *ZDM Mathematics Education, 39*, 79–86. https://doi.org/10.1007/s11858-006-0006-z

Kieran, C. (2006). Research on the learning and teaching of algebra: A broadening of sources of meaning. In A. Gutiérrez & P. Boero (Eds.), *Handbook of research on the psychology of mathematics education* (pp. 11–50). Sense.

Kilpatrick, J., Hoyles, C., & Skovsmose, O. (2005). *Meaning in mathematics education*. Kluwer.

Lannin, J. K. (2005). Generalization and justification: The challenge of introducing algebraic reasoning through patterning activities, *Mathematical Thinking and Learning, 7*(3), 231–258. https://doi.org/10.1207/s15327833mtl0703_3

Lesh, R., & Zawojewski, J. (2007). Problem solving and modeling. In F. K. Lester Jr. (Ed.), *Second handbook of research on mathematics teaching and learning* (pp. 763–804). Information Age Publishing.

Lester, F. K., Jr. (2013). Thoughts about research on mathematical problem-solving instruction. *The Mathematics Enthusiast, 10*(1–2), 245–278. https://scholarworks.umt.edu/tme/vol10/iss1/12

Lüken M. M. (2020). Patterning as a mathematical activity: An analysis of young children's strategies when working with repeating patterns. In M. Carlsen, I. Erfjord, & P. Hundeland (Eds.), *Mathematics education in the early years* (pp. 79–92). Springer. https://doi.org/10.1007/978-3-030-34776-5_5

Maglipong, C. V., & Bongolto, J. L. (2017). Mathematical patterns: It's impact on students' performance in college algebra. *International Journal of Science and Research, 6*(1), 2112–2119. https://www.ijsr.net/search_index_results_paperid.php?id=20011701

Mamona-Downs, J., & Downs, M. (2013). Problem solving and its elements in forming proof. *The Mathematics Enthusiast, 10*(1–2), 137–162. https://scholarworks.umt.edu/tme/vol10/iss1/12

Masingila, J. O., Lester, F. K., & Raymond, A. M. (2011). *Mathematics for elementary teachers via problem solving: Instructor resource manual* (3rd ed.). Copley Custom Textbooks.

Masingila, J. O., Olanoff, D., & Kimani, P. M. (2018). Mathematical knowledge for teaching teachers: Knowledge used and developed by mathematics teacher educators in learning to teach via problem solving. *Journal of Mathematics Teacher Education, 21*, 429–450. https://doi.org/10.1007/s10857-017-9389-8

Markworth, K. A. (2016). A repeat look at repeating patterns. *Teaching Children Mathematics, 23*(1), 22–29. https://doi.org/10.5951/teacchilmath.23.1.0022

Masingila, J. O., Olanoff, D., & Kimani, P. M. (2018). Mathematical knowledge for teaching teachers: Knowledge used and developed by mathematics teacher educators in learning to teach via problem solving. *Journal of Mathematics Teacher Education, 21*, 429–450. https://doi.org/10.1007/s10857-017-9389-8

Michael, S., Elia, I., Gagatsis, A., Theoklitou, A., & Savva, A. (2006). Levels of understanding of patterns in multiple representations. In J. Novotná, H. Moraová, M. Krátká, & N. Stehlíková (Eds.), *Proceedings 30th Conference of the International Group for the Psychology of Mathematics Education* (Vol. 4, pp. 161–168). PME.

Mulligan, J., & Mitchelmore, M. (2009). Awareness of pattern and structure in early mathematical development. *Mathematics Education Research Journal, 21*(2), 33–49. https://doi.org/10.1007/BF03217544

National Research Council (NRC). (1989). *Everybody counts: A Report to the nation on the future of mathematics education*. The National Academies Press. https://doi.org/10.17226/1199.

National Research Council (NRC). (1990). *Reshaping school mathematics: A philosophy and framework for curriculum*. National Academy Press. https://doi.org/10.17226/1498

National Research Council (NRC). 2001. *Adding it up: Helping children learn mathematics*. The National Academies Press. https://doi.org/10.17226/9822.

Noss, R., Healy, L., & Hoyles, C. (1997). The construction of mathematical meanings: Connecting the visual with the symbolic. *Educational Studies in Mathematics 33*(2), 203–233. https://doi.org/10.1023/A:1002943821419

Oliveri, G. (1997). Mathematics. A science of patterns? *Synthese, 112*(3), 379–402. https://doi.org/10.1023/A:1004906107430

Ohlsson, S. (1993). Abstract schemas. *Educational Psychologist, 28*(1), 51–66. https://doi.org/10.1207/s15326985ep2801_5

Orton, A., & Orton, J. (1999). Pattern and the approach to algebra. In A. Orton (Ed.). *Pattern in the teaching and learning of mathematics* (pp. 104–120). Cassel.

Orton, A., Orton, J., & Roper, T. (1999). Pictorial and practical contexts and the perception of pattern. In A. Orton (Ed.), *Pattern in the teaching and learning of mathematics*. Cassell.

Pitta-Pantazi, D., Gray, E., & Christou, C. (2004). Elementary school students' mental representations of fractions. In M. J Høines & A. Fuglestad (Eds.), *Proceedings of the 28th annual conference of the International Group for the Psychology of Mathematics Education* (Vol. 4, pp. 41–48). Bergen University College.

Pólya, G. (1954). *Mathematics and plausible reasoning* (Volume 1, Induction and analogy in mathematics; Volume 2, Patterns of plausible inference). Princeton University Press.

Pólya, G. (1957). *How to solve it* (2nd ed.). Princeton University Press.

Prediger, S. (2010). How to develop mathematics-for-teaching and for understanding: The case of meanings of the equal sign. *Journal of Mathematics Teacher Education, 13*(1), 73–93. https://doi.org/10.1007/s10857-009-9119-y

Resnik, M. D. (1981). Mathematics as a science of patterns: Ontology and reference. *Noûs, 15*(4), 529–550. https://doi.org/10.2307/2214851

Resnik, M. D. (1982). Mathematics as a science of patterns: Epistemology. *Noûs, 16*(1), 95–105. https://doi.org/10.2307/2215419

Rivera, F. D., & Becker, J. R. (2005). Teacher to teacher: Figural and numerical modes of generalizing in algebra. *Mathematics Teaching in the Middle School, 11*(4), 198–203. https://doi.org/10.5951/MTMS.11.4.0198

Schoenfeld, A. H. (1992). Learning to think mathematically: Problem solving, metacognition, and sense-making in mathematics. In D. Grouws (Ed.), *Handbook of research on mathematics teaching and learning* (pp. 334–370). Macmillan.

Stacey, K. (1989). Finding and using patterns in linear generalising problems. *Educational Studies in Mathematics, 20*(2), 147–164. https://doi.org/10.1007/BF00579460

Steen, L. A. (1988). The science of patterns. *Science, 24*, 611–616. https://doi.org/10.1126/science.240.4852.611

Stein, M. K., Boaler, J., & Silver, E. A. (2003). Teaching mathematics through problem solving: Research perspectives. In H. L. Schoen & R. I. Charles (Eds.), *Teaching mathematics through problem solving: Grades 6–12*. National Council of Teachers of Mathematics.

Stylianides, G. J., & Silver, E. (2009). Reasoning-and-proving in school mathematics: The case of pattern identification. In D. A. Stylianou, M. L. Blanton, & E. J. Knuth (Eds.), *Teaching and learning proof across the grades: A K–16 perspective* (pp. 235–249). Routledge.

Tsamir, P., Tirosh, D., Levenson, E. S., Barkai, R., & Tabach, M. (2017). Repeating patterns in kindergarten: Findings from children's enactments of two activities. *Educational Studies in Mathematics, 96*(1), 83–99. https://doi.org/10.1007/s10649-017-9762-7

Vale, I., & Cabrita, I. (2008). Learning through patterns: A powerful approach to algebraic thinking. In K. Kumpulainen & A. Toom (Eds.), *The Proceedings of the 18th Annual Conference of the European Teacher Education Network* (pp. 63–69). Published Jointly by ETEN (The European Teacher Education Network), University of Helsinki, Department of Applied Sciences of Education, and CICERO Learning, University of Helsinki.

Vanluydt, E., Wijns, N., Torbeyns, J., & VanDooren, W. (2021). Early childhood mathematical development: The association between patterning and proportional reasoning. *Educational Studies in Mathematics, 107*, 93–110. https://doi.org/10.1007/s10649-020-10017-w

Vogel, R. (2005). Patterns—A fundamental idea of mathematical thinking and learning. *ZDM—Mathematics Education, 37*(5), 445–449. https://doi.org/10.1007/s11858-005-0035-z

Warren, E. (2005). Young children's ability to generalise the pattern rule for growing patterns. In H. Chick & J. Vincent (Eds.), *Proceedings of the 29th conference of the International Group for the Psychology of Mathematics Education* (Vol. 4, pp. 305–312). PME.

Warren, E. A., Cooper, T. J., & Lamb, J. T. (2006). Investigating functional thinking in the elementary classroom: Foundations of early algebraic reasoning. *Journal of Mathematical Behavior, 25*, 208–223. https://doi.org/10.1016/j.jmathb.2006.09.006

Wittmann, E. C. (2005). *Mathematics as the science of patterns—A guideline for developing mathematics education from early childhood to adulthood.* Plenary Lecture presented at the International Colloquium "Mathematical Learning from Early Childhood to Adulthood" organized by the Centre de Recherche sur l'Enseignement des Mathematiques in collaboration with the Institut de mathématique de l'Université de Mons-Hainaut, Mons/Belgium, July 7–9.

Zazkis, R., & Liljedahl, P. (2002). Generalization of patterns: The tension between algebraic thinking and algebraic notation. *Educational Studies in Mathematics, 49*(3), 379–402. https://doi.org/10.1023/A:1020291317178

Zopf, D. (2010). *Mathematical knowledge for teaching teachers: The mathematical work of and knowledge entailed by teacher education* [Unpublished doctoral dissertation]. http://deepblue.lib.umich.edu/bitstream/handle/2027.42/77702/dzopf_1.pdf

CHAPTER 2

PÓLYA REVISITED

The Development and Implementation of a Problem-Solving Self-Efficacy Instrument With Preservice Elementary Mathematics Teachers

James A. Telese and Jair J. Aguilar

INTRODUCTION

The belief that an individual has in regard to their own ability to perform or complete a task is the "I can" or "I can't" belief, which impacts how a person decides to solve a problem or resolve a problem-solving situation (Siegle & McCoach, 2007). In relation to mathematics learning, students' self-efficacy plays an important role in how they approach a problem or task, and how confident and persistent they feel in order to find a solution or to complete a complex task (Pajares, 1996).

This is commonly known as self-efficacy. The theory of self-efficacy was developed by Bandura (1986). According to Bandura (1994), efficacy is a belief in one's personal capabilities. The theory suggests that self-efficacy is the most influential aspect of one's self-knowledge, which fosters the relationship between knowledge and action. Those who have high self-efficacy expect favorable outcomes, and those who doubt themselves may

Mathematics as the Science of Patterns:
Making the Invisible Visible to Students Through Teaching, pp. 17–39

experience negative outcomes (Bandura, 1986). There are four main influences on the degrees of self-efficacy (Bandura 1997), which may be experienced through professional development experiences: (1) *Mastery experiences*—a person achieves success and as a result becomes more confident in his or her abilities; (2) *Vicarious experiences provided by social models*—a person strengthens his or her self-beliefs by observing someone similar who finds success; (3) *Social persuasion*—a person is encouraged or verbally persuaded that he or she possesses the capabilities to master a given activity; and (4) *Stress reduction*—a person's negative emotional state is altered to adjust his or her judgment of personal self-efficacy.

Teachers with a high sense of instruction efficacy believe that difficult students are teachable through extra effort and appropriate techniques, whereas teachers who have a low sense of instructional efficacy believe that there is little they can do to motivate the student (Bandura, 1997). The current study focused on mathematics teacher self-efficacy for problem solving. We used a modified version of a problem-solving self-efficacy instrument (Heppner & Petersen, 1982) to measure preservice teacher's belief before and after completing a series of non-routine problem-solving tasks. The purpose of the study was to measure elementary preservice teachers' self-efficacy toward mathematical problem solving when they are exposed to nonroutine tasks.

PERSPECTIVES

Mathematics teaching and preservice elementary teacher preparation with emphasis on teacher's self-efficacy toward mathematical problem solving is the primary focus of this study. Perspectives examined in this section focus on the meaning of problem solving, problem-solving self-efficacy, problem solving processes, and preservice teacher attitudes toward mathematics.

Elementary Teacher Preparation and Problem Solving

Organizations such as the National Council of Teachers of Mathematics (1980, 2000) have long held the view that mathematics should be a sense making activity stemming from problem solving. Educators, parents, and children see differently the meaning of problem solving (Feikes et al., 2018). Generally, it may be considered engaging students in challenging, nonroutine problems. Another aspect relates to teaching practices by taking a problem-solving approach characterized by small group work and whole class discussions (Feikes et al., 2018). This section describes the importance of preparing elementary teachers to take this problem solving or inquiry-based approach for developing mathematics understanding.

The National Academies of Sciences, Engineering, and Medicine (2020) contended that a school's success hinges on the effectiveness of teachers in the classroom. There is a fair amount of certainty based on research conducted over the last 20 years that teachers play a critical role in the success of their students, both academically and with respect to longer term outcomes (National Academies of Sciences, Engineering, and Medicine, 2020). Preservice and in-service teachers tend to have low self-efficacy toward mathematics which hinders them to teach mathematics concepts. Consequently, it is likely that they will not promote the development of problem-solving skills with their students (Starko & Schack,1989). It is necessary that preservice elementary teachers develop deep and connected understandings of important mathematical ideas (Ball & Bass, 2000; Ma, 1999) during their teacher preparation program. Elementary teacher preparation influences the way mathematics is taught by beginning teachers and the frequency of mathematics lessons that are presented (Gimbert et al., 2007).

Olanoff et al. (2009) proposed that one way to foster deep mathematical understanding in preservice elementary school teachers is to engage them in learning mathematics through problem solving. Problem solving has long been viewed as an important aspect of learning mathematics. The National Council of Teachers of Mathematics (NCTM) encouraged problem solving in K–12 mathematics instruction since the 1980s with the publication of *An Agenda for Action* (NCTM, 1980) that focused heavily on problem solving, to note the first recommendation, "The mathematics curriculum should be organized around problem solving" (NCTM, 1980 p. 2). This meant that mathematics programs should be designed to "equip" students with strategic competency, adaptive reasoning, and productive dispositions (National Research Council, 2001). The goal of *An Agenda for Action* was to encourage a mathematics curriculum, "at all grade levels" organized around culturally relevant problem solving, so that the students could learn to deal with daily life experiences, whether expected or unexpected, and to solve real-world problems (NCTM, 1980).

More recently, the National Council for Teacher of Mathematics (2014) encouraged teachers to implement tasks that promote reasoning and problem-solving as well as to support productive struggle while learning mathematics. This may be difficult for some teachers to implement if their self-efficacy for mathematics teaching and problem solving is low, having little confidence to solve problems and to teach mathematics is often associated with elementary teachers (Ma, 2020). One approach to mathematics teaching is teaching-through-problem solving where the development of problem-solving skills is not separated from domain-specific objectives

(Cai, 2010). Teaching through problem solving starts with exploring a mathematically rich problem from which students learn and understand important aspects of mathematics concepts. The tasks should be open-ended and allow for multiple solution pathways and may have multiple correct answers. There is collective sharing and discussion of different attempts and solutions that help students to evaluate their and others solution processes (Cai, 2010). Hence, elementary teachers need to have the ability and be confident in their ability to teach through problem solving.

Problem Solving Self-Efficacy

Self-efficacy plays an important role in whether an individual will engage in problem solving or persist when solving a nonroutine problem (Schoenfeld, 1985, 2013). Several researchers (e.g., Briley, 2012; Pendergast et al., 2011) have measured self-efficacy of preservice elementary teachers during their preparation programs. Bal (2014) used an adapted mathematical problem-solving belief-scale that was originally developed by Kloosterman and Stage (1992). However, this instrument does not directly measure confidence in the process of problem solving. Ramirez et al. (2016) examined the relationship between problem solving strategies and mathematics anxiety of early elementary students. They concluded that mathematics anxiety predicts children's reduced use of advanced problem-solving strategies. Mathematical problem-solving beliefs were included in a study that examined their role in academic performance of middle school students (Schommer-Aikins et al., 2005). The researchers used the Indiana Mathematics Scale (Kloosterman & Stage, 1992) to assess beliefs about problem solving. One strategy that was used was to determine whether general and mathematical beliefs predicted mathematical problem-solving beliefs. Problem solving and problem posing have drawn international attention. In a recent publication, Felmer et al. (2016) examined various aspects of problem solving and problem posing as a result of an international, comparative research project by universities in Helsinke and Santiago, Chile. In one study, Carrillo and Cruz (2016) used a questionnaire to examine two 14 year old subjects problem solving ability. Preservice elementary mathematics teachers' problem-solving strategies were examined by Avcu and Avcu (2010). They found that their use of different strategies to solve problems was limited. Chapman (2013), based on a review of literature, contended that teachers need to have knowledge of mathematical problem solving for themselves as problem solvers and to help their students become problem solvers too.

It is highly regarded that the initial preparation of teachers should focus on teachers' knowledge of the subject, on the curriculum, or on how students learn the subject (cf. Ball & Forzani, 2011; Philipp et al., 2007). The Council for the Accreditation of Educator Preparation (2013) in their publication, *Accreditation Standards and Evidence: Aspirations for Educator Preparation* indicated that teacher candidates must learn "critical concepts and principles of their discipline and, by completion, are able to use discipline-specific practices flexibly" (p. 2). One of these specific practices is problem solving. Education Northwest (2011) recommended a problem-centered approach to elementary mathematics teaching, suggesting that this approach helps children develop understanding of mathematical concepts. This encourages the development of new strategies as they encounter various types of problems and tasks.

Problem solving at the elementary grade levels can contribute to the students' confidence to do mathematics and achievement (Jitendra et al., 2005). Elementary teachers must themselves be confident in teaching mathematics and understand the importance of problem solving. Elementary teacher preparation programs, including their field experiences, influence preservice teachers' confidence and ability for teaching mathematics (Kirtman, 2008). For this reason, reform in elementary mathematics teacher preparation should challenge preservice teachers' perceptions of mathematics as rules to follow and memorization (Budak & Budak, 2011), and mathematics teaching as telling (Ma, 2020), to one that focuses on culturally relevant problem solving (Leonard et al., 2010).

Problem Solving Processes

Problem solving may be taught through the use of various heuristics by developing a repertoire of problem-solving strategies introduced by the teacher (Feikes et al., 2018). Teaching mathematics with a problem-solving focus helps to enhance understanding of mathematics concepts (Schoenfeld, 1985). In turn, this would foster the ability to consider different perspectives and solution pathways, and to think creatively rather than an emphasis on procedural skills, where problems have a one-way, or a memorized algorithmic approach to the solution. Schoenfeld (1985) proposed four categories of knowledge and skills that are needed to be successful when solving mathematics problems. The first are the resources, which refers to the plans and procedures of mathematics. The second, heuristics. It refers to the strategies and techniques that an individual needs to solve a problem successfully. The third category is control, which refers to decision making while the problem is being solved, ultimately deciding on the best way to solve the problem, including what resources will be needed and what strategies should be followed. Fourth, Schoenfeld (1985, 2013) considers the role of an individual's belief about themselves as problem solvers. A person's belief system develops overtime and is influenced by their background and experiences which in turn influences the way a problem is solved and what approaches are taken.

Representation of mathematics problems play a role when solving problems. Lesh and Doerr (2003) contended that problem solving should encourage the use of models and modeling. Lesh and Doeer described mathematical modeling as that process where conceptual systems and models are to be used to create and develop new models that could serve as a possible solution to different contexts and situations. For Lesh and Doeer, a model represents "the product" that evolved from a real situation, and modeling the process followed to create that model (Sriraman & Lesh, 2006). Modeling requires the solver to transfer a real situation into a mathematical expression. A model is developed and tested with the intent to provide a solution for the real-world problem situation. The solver should repeat this process as many times as necessary until a tested model is developed. The development of a mathematical model stimulates reflection on the experiences that occurred during problem solving (Lesh & Zawojewski, 2007, p. 770), which also lead the solver to consider different perspectives.

In a more heuristic approach to problem solving, which refers to solving a task in a systematic way (Koichu et al., 2004), Guzman (2018) stated that students should engage in problem-solving using a three-step approach: Rehearsing, Elaborating, and Organizing. Rehearsing refers to understanding the task, reading it as many times as necessary, and having a clear comprehension of what the problem is requiring the solver to do. In addition, this step requires the problem solver to use content and background knowledge that could potentially help in determining a solution pathway. Elaborating involves close consideration of the details of the problem by underlining and highlighting aspects and words that prompt questions such as, what is the problem? and how is it to be solved? which refers to how the problem solver may organize and structure the provided information in order to facilitate a solution.

A famous problem-solving heuristic was introduced by Pólya (1945) known as Pólya's 4-Step Problem-Solving Process. Pólya's aims were to devise a structured framework so that teachers could implement it with their students, and so that students could use a method to find a solution to a mathematical task (Chapman, 2013). Pólya's 4-steps are well known: (1) Understand the problem, (2) Devise a plan, (3) Carry out the plan, and (4) Look back. Understanding the problem means for the problem solver to have a full understanding of what the problem is about, and what elements need to be considered to arrive at a solution. This phase involves teachers in orchestrating a question discussion session to confirm that the students have complete understanding of the situation stated in the problem. Some of the questions Pólya suggested included, what word or words are not understood in the problem? What is the problem asking us to do? Can you restate the problem in your own words? Similar to Lesh and Doerr's (2003) modeling perspective, could you draw a picture to represent the problem?

The second step asks both students and teachers to create a strategy that would help them solve the problem. In an open-ended task there may be more than one solution path. Pólya (1945) suggested that students should pick one that they think will be the most appropriate approach. Some of the techniques suggested involving the elimination of possibilities, direct reasoning, modeling, finding patterns, working backward, or using a formula.

The third step, carry Out the Plan, requires students to be persistent with the chosen strategy, since the chosen approach might not be workable, leading to the development of another plan. This cyclic process is what Lesh and Doeer (2003) called the modeling cycle, where different perspectives of the solution emerge while solving a task.

The fourth step of Pólya's (1945) problem solving process is to revise what has been done, reflect on the work completed, and gain insight on how the solution-process could be used in other similar problem-solving situations. This builds a repertoire of strategies that may be used with future problem-solving tasks.

The problem-solving process developed by Pólya (1945), is intended to help students successfully connect mathematical ideas (Hiebert et al., 1997; Lambdin, 2003), and develop the ability to successfully solve problems (NCTM, 2000). Pólya's strategies may be useful with a variety of nonroutine tasks including real-life problem solving (NCTM, 2000; Lesh & Zawojewski, 2007). In this context, problem-solving tasks (Kolovou et al., 2009) were implemented in this study we are describing here, and where preservice teachers depict their self-efficacy and belief toward mathematics before and after the implementation of the task. We considered Pólya's (1985) heuristic framework as a lens for the analysis.

Brunn (2013) conducted a qualitative study with 70 in-service elementary teachers in grades two to five to examine what strategies the teachers reported using to teach solving of word problems. Data was collected through 70 interviews. The interview schedule consisted of 12 open-ended questions. Predetermined categories were established based on NCTM Standards (2000). The results showed that the 40% reported teaching Drawing a Picture to solve a word problem, followed by 20% who reported teaching students to Make a table or graph.

Attitudes Toward Mathematics

Similar to the role of self-efficacy for problem solving is an individual's attitudes toward mathematics that have developed over time. One's own ideas and experiences help to form a mindset that guides our decision-making processes, and consequently creates an attitude toward someone or something (Awofala, 2016). In learning mathematics, students' attitude is considered to be a cause for them to see mathematics as something positive or negative that could potentially trigger a certain way of thinking or behaving (Jovanovic & King, 1998). Neale (1969) mentioned that a student's attitude toward mathematics is "the liking or disliking of mathematics, a tendency to engage in or avoid mathematical activities, a belief that one is good or bad at mathematics, and a belief that mathematics is useful or useless" (p. 632). Neale's statement is convincing because when its extrapolated to students in any context or level, it can be proven that the attitude makes a difference among students regardless of how challenging the class is perceived. When the student's attitude is positive, it's likely that the student would be more motivated to learn, advance, and potentially incline to follow a STEM career path. In contrast, when the attitude toward mathematics is negative, the student would potentially have a low confidence, motivation, and engagement toward enhancing anything related to mathematics (Aljaberi, 2014; Asante, 2012).

Attitudes toward mathematics develop over time and change as well over time. Elementary students often have a positive attitude toward mathematics, and generally do well in mathematics tasks (Yilmaz et al., 2010). Providing positive mathematical experiences at the elementary level sets the stage for academic success. Lawsha and Hussain (2011) identified three main elements that have an effect on a student's attitude toward mathematics: (1) the school, the teacher and the type of teaching, (2) the student's attitude, and (3) the social and environmental context. In regard to the school, the teacher, and the type of teaching, involves presenting lessons that are motivating and engaging, classroom management and discipline, and the rules the school have potential to impact students' attitudes toward mathematics and schooling (Awofala, 2016; Cater & Norwood, 1997). Similarly, the students' degree of self-confidence, their anxieties, their self-efficacy and past experiences, and parental involvement or expectations (Köğce et al., 2009) also impacts their attitude (Klein, 2004).

The teacher and the teaching style could be considered the one factor with the highest potential to impact student attitudes, particularly for elementary teachers (Ball & Forzani, 2011; Schmoker, 2006). A teacher with a negative attitude toward mathematics, will likely project that attitude to the students, hampering and blocking the student's true potential. Therefore, it is vital that preservice teachers are provided with experiences that enhance their mathematical pedagogical and content knowledge in ways that help them support and promote mathematics understanding among their future young students (Nuangchalerm, 2012).

Although teacher preparation programs require their students to enroll and pass mathematics courses such as college algebra or fundamentals of mathematics, many elementary preservice teachers still lack mathematical content knowledge. Content knowledge is necessary in order to engage students in deep understandings of mathematical concepts (Association of Mathematics Teacher Educators [AMTE], 2017). Elementary teachers are notorious for having low self-efficacy for teaching mathematics especially when working on problems that do not have a clear solution pathway (Temel, 2014).

This poor perception of mathematics and of themselves as mathematics teachers negatively influences preservice teachers' attitudes and beliefs toward problem solving (Gresham, 2008). Other influences on the formation of their attitudes toward mathematics include their previous experiences in mathematics courses (Bekdemir, 2010), or a lack of content knowledge (Swars et al., 2007). If preservice teachers complete their teacher preparation program with all or some of the above experiences, it is likely they will not promote problem solving within their students. To change preservice teachers' attitudes toward mathematics problem solving, they need to be exposed and challenged to solve problems that enhance their confidence, self-efficacy, and attitude towards mathematics (Wilson & Cooney, 2002). As Swars et al. (2007) stated, "this is a challenging process that requires "thoughtful reflection and examination of teaching and learning" (p. 325).

METHODOLOGY

An exploratory factor analysis was conducted, from which four main components emerged: Confidence in Understanding the Problem, Confidence in Devising a Plan, Confidence in Carrying Out the Plan, and Confidence in Reflecting on the Strategy. In addition, a paired *t*-test showed no statistically significant differences in confidence levels from the beginning of the semester and near the end.

Participants

The study was conducted in a large Hispanic Serving Institution located in deep South Texas from the Fall semester of 2018 to the Fall Semester of 2019. In total there were 213 preservice elementary teachers enrolled in an elementary mathematics methods course. The sample size consisted of 200 preservice teachers, due to missing data, which were used for the analysis. There were 182 females (91%), and 18 males (9%). The majority of the participants were of Mexican or Mexican American descent.

Treatment

The elementary mathematics course, along with the textbook topics presented in *Teaching Elementary and Middle School Mathematics* by Van de Walle et al. (2019), included a supplemental focus on the development of problem-solving skills. The preservice teachers were provided non-routine tasks with the goal of building their self-efficacy and confidence for problem solving. The tasks were implemented every other week during the 14-week semester. The preservice teachers were required to solve the problems on their own, and then return to class with their solution. Working in cooperative groups, solutions were compared with each other. The instructor then led a class discussion over selected versions to demonstrate various approaches used by the students. They worked on a total of nine non-routine mathematical tasks.

Data Collection

A survey developed by Heppner and Peterson (1982) that measures dispositions and beliefs toward problem-solving was modified (see Appendix A) to align with the mathematics problem solving. It was administered over three consecutives semesters. The survey was administered at the beginning of the semester as a pretest and toward the end of the semester as a posttest. The survey had 32 statements with a 5-point Likert Scale where 1 represented Never, 2-Almost Never, 3-Occasionally, 4-Almost every time, and 5-Every time. An example of an item is "I trust my ability to solve new and difficult problems." Another item is "When I work on an open-ended problem, I feel like I do not know what to do to solve it."

RESULTS

Exploratory Factor Analysis

An exploratory factor analysis was conducted using the posttest survey responses. A KMO and Bartlett's Test for determining the adequacy of the sample indicated that the sample size was adequate. Table 2.1 presents the results of the KMO and Bartlett's Test. The Principal Component with Varimax Rotation analysis revealed at least 4 factors having eigen-values

Table 2.1

Results of the KMO and Bartlett's Test of Sphericity

Kaiser-Meyer-Olkin Measure of Sampling Adequacy	.790
Bartlett's Test of Sphericity	
Approx. Chi-Square	1804.880
df	496

greater than 1.00 and accounting for 50% of the common variance. Figure 2.1 presents the Scree Plot indicating that at the inflection point there are four factors. We decided to run the factor analysis to force a four-factor analysis. The 4 components produced 50% of the variance. In addition, to test for the instrument's reliability, a Cronbach Alpha was determined to be 0.600 for the 32 items in the survey. Figure 2.1 displays a Scree Plot revealing four factors. A four-factor rotation was conducted and the four factor rotated component matrix is presented in Table 2.2.

Figure 2.1

Scree Plot Showing Four Factors

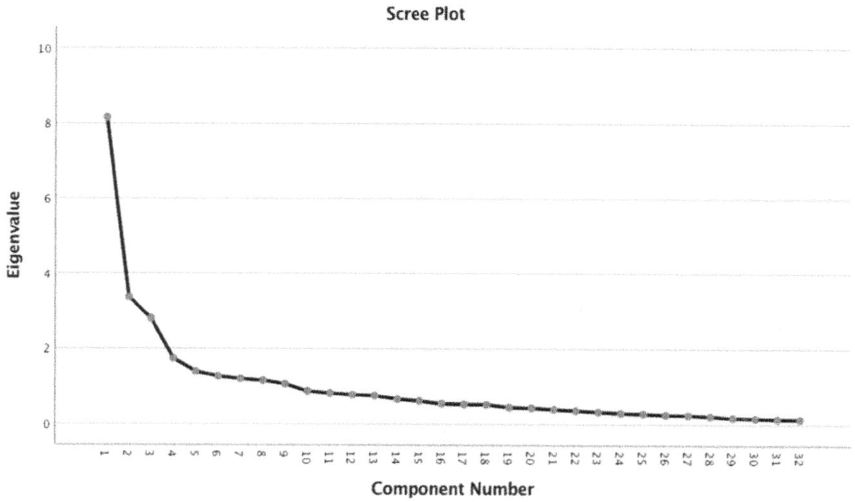

Table 2.2

Four Factor Rotated Component Matrix

ITEM	Component			
	1	2	3	4
8. When confronted with an open-ended problem (problem with no obvious solution or strategy), I am able to determine what the problem requires for a solution.	.605			
9. I have the ability to solve most open-ended problems (problem with no obvious solution or strategy) even though initially no solution is immediately apparent.	.666			
11. As I work on a solution to an open-ended problem (problem with no obvious solution or strategy), I make decisions on how to proceed and I am happy with them later.	.553			
18. When I make a plan to solve an open-ended problem (problem with no obvious solution or strategy), I am almost certain that I can make it work.	.678			
21. Given enough time and effort, I believe I can solve most open-ended problems that confront me.	.614			
22. When faced with an open-ended problem that I don't understand immediately, I have confidence I can eventually solve it.	.592			
25. I trust my ability to solve new and difficult problems.	.545			
30. After making a decision on how to solve an open-ended problem (problem with no obvious solution or strategy), the outcome I expected usually matches the actual outcome.	.587			
6. After I have tried to solve a open-ended problem with a specific strategy, I take time and compare the answer with what I had predicted the answer should be.		.721		
7. When I have a problem, I think up as many possible ways to solve it as I can until I can't come up with any more ideas.		.731		
15. When confronted with an open-ended problem (problem with no obvious solution or strategy), I stop and think about it before deciding on a next step.		.549		

(Table contined on next page)

Table 2.2 (Continued)

Four Factor Rotated Component Matrix

ITEM	Component			
	1	2	3	4
17. When deciding on how to solve an open-ended problem (problem with no obvious solution or strategy), I weigh the efficiency of each strategy and compare them against each other.		.707		
19. I try to predict the overall result of carrying out a particular course of actions.		.531		
26. I compare possible solutions to make a decision on which strategy to use.		.699		
1. When a solution to an open-ended mathematics problem (with no obvious solution or strategy) was unsuccessful, I do not reflect on why it didn't work.			.705	
2. When I am confronted with a complex open-ended mathematics problem (problem with no obvious solution or strategy), I do not bother to think about a strategy to solve the problem in relation to what is being asked.			.766	
3. When my first efforts to solve a problem fail, I become anxious about my ability to use a different strategy to solve the problem.			.618	
4. After I have solved an open-ended problem (problem with no obvious solution or strategy), I do not analyze what went right or what went wrong.			.623	
10. Many open-ended problems (problem with no obvious solution or strategy) I solve are too complex for me to solve.			.429	
13. Sometimes when I try to solve open-ended problems (problem with no obvious solution or strategy), I do not stop and take time to think about the problem, but just keep on working on it.			.535	
14. When deciding on a possible solution to an open-ended problem (problem with no obvious solution or strategy), I do not take time to consider other possible solutions.			.463	

(Table contined on next page)

Table 2.2 (Continued)

Four Factor Rotated Component Matrix

ITEM	Component			
	1	2	3	4
20. When I try to think up possible solutions to a open-ended problem, I do not come up with very many alternatives.			.456	
23. When I work on an open-ended problem (problem with no obvious solution or strategy), I feel like I do not know what to do to solve it.			.495	
16. I generally go with the first good idea that comes to my mind.				.640
28. When I am confused by an open-ended problem (problem with no obvious solution or strategy), one of the first things I do is analyze what the problem requires and what I may need to do.				.491
29. Sometimes I get confused, which keeps me from considering other ways of solving the problem.				.400
32. When I first encounter an open-ended problem (problem with no obvious solution or strategy), one of the first things I do is to determine relevant information needed to solve the problem.				.606

Paired *t*-test

Based on the factor loadings, the factors that were identified aligned with Pólya's Four-step problem solving process. The four factors are Confidence in Understanding the Problem, Confidence in Devising a Plan, Confidence in Carrying Out the Plan, and Confidence in Reflecting on the Strategy. The items that loaded in each of the components were summed to create the four factors for running a t-test analysis. Items 5 and 12 were omitted from the summing process due to their loading on more than one factor. Table 2.3 presents the means and standard deviations for the prepost pairs for each of the four components. The results of a paired *t*-test showed no statistically significant differences in confidence levels from the beginning of the semester and near the end of the semester.

Table 2.3

Paired Component Means and Standard Deviations

	Component	Mean	*N*	SD	Std. Error Mean
Pair 1	Pre-Confidence in Understanding the Problem	28.823	193	4.090	.29441
	Post-Confidence in Understanding the Problem	29.171	193	3.879	.27924
Pair 2	Pre-Confidence in Devising a Plan	20.015	196	3.745	.26750
	Post-Confidence in Developing a Plan	20.301	196	4.264	.30456
Pair 3	Pre-Confidence in Carrying Out the Plan	21.778	194	4.702	.33760
	Post-Confidence in Carrying Out the Plan	21.551	194	5.045	.36220
Pair 4	Pre-Confidence on Reflecting on Strategy	14.775	200	1.874	.13251
	Post-Confidence in Reflecting on Strategy	14.580	200	1.942	.13734

DISCUSSION

The survey results showed that the items were consistent with tenets from Pólya's four-step approach. The analysis revealed that the survey was responded to in a consistent manner. The reliability coefficient indicated that the survey has internal consistency and would offer consistent results. The exploratory factor analysis supported a four-component survey aligned with Pólya's four-step problem solving solution. Component 1 was determined to be Confidence in Understanding the Problem which consisted of eight items. Component 2 was deemed Confidence in Developing a Plan to solve the problem which consisted of six items. Component 3 was aligned with Confidence in Carrying Out a Plan for problem solving which consisted of nine items. Component Four was interpreted to be Confidence in Reflection on the Strategy which consisted of four items.

The results of the paired *t*-test were disconcerting because it was thought that exposure to problem solving tasks would foster greater confidence in problem solving. The results indicated that the preservice teachers held moderate levels of problem-solving confidence as they entered the course as when they were completing the course. Further research is needed to explore confidence levels through a mixed methods approach that may include clinical interviews and a greater frequency of administering problem solving tasks. Another possible explanation may be that the preservice teachers did not take the tasks and the survey seriously.

CONCLUSION

Through a mixed methods approach that may include interviews with participants that may reveal further insights into their confidence to develop mathematics understanding through problem solving. However, the recommendation of exposing preservice teachers to problem solving is one that should be heeded even though the current study did not support the notion that solving open-ended mathematics problems regularly will help develop confidence. Perhaps a more focused approach on problem solving rather than used as an additional topic of the course could have positive approach.

Telese et al. (2020) showed that teacher self-efficacy played a role in promoting student achievement, and Telese (2016) found that in-service elementary mathematics teachers' self-efficacy improved after experiencing two semesters of weekly Saturday, eight-hour sessions that focused on developing content and pedagogical content knowledge. This suggests that it may take an extended period of time or at least, longer than one semester to effect change in problem solving confidence.

The current study showed that preservice elementary teachers' gained confidence in coming to understand the problems, which is an important first step when engaged in nonroutine tasks. Furthermore, the implications of the study suggest that preservice elementary teachers could benefit from a more focused approach to learning to teach mathematics through problem solving in their preparation program. Another avenue for further research would be to observe those students who experience regular difficulties in solving nonroutine mathematical tasks.

APENDIX A

Instrument

1. When a solution to an open-ended mathematics problem (with no obvious solution or strategy) was unsuccessful, I do not reflect on why it didn't work.

2. When I am confronted with a complex open-ended mathematics problem (problem with no obvious solution or strategy), I do not bother to think about a strategy to solve the problem in relation to what is being asked.
3. When my first efforts to solve a problem fail, I become anxious about my ability to use a different strategy to solve the problem.
4. After I have solved an open-ended problem (problem with no obvious solution or strategy), I do not analyze what went right or what went wrong.
5. I am usually able to think up effective alternatives (i.e., appropriate ways or strategies) to solve an open-ended problem (problem with no obvious solution or strategy).
6. After I have tried to solve an open-ended problem with a specific strategy, I take time and compare the answer with what I had predicted the answer should be.
7. When I have a problem, I think up as many possible ways to solve it as I can until I can't come up with any more ideas.
8. When confronted with an open-ended problem (problem with no obvious solution or strategy), I am able to determine what the problem requires for a solution.
9. I have the ability to solve most open-ended problems (problem with no obvious solution or strategy) even though initially no solution is immediately apparent.
10. Many open-ended problems (problem with no obvious solution or strategy) I solve are too complex for me to solve.
11. As I work on a solution to an open-ended problem (problem with no obvious solution or strategy), I make decisions on how to proceed and I am happy with them later.
12. When confronted with an open-ended problem (problem with no obvious solution or strategy), I tend to do the first thing that I can think of to solve it.
13. Sometimes when I try to solve open-ended problems (problem with no obvious solution or strategy), I do not stop and take time to think about the problem, but just keep on working on it.
14. When deciding on a possible solution to an open-ended problem (problem with no obvious solution or strategy), I do not take time to consider other possible solutions.
15. When confronted with an open-ended problem (problem with no obvious solution or strategy), I stop and think about it before deciding on a next step.
16. I generally go with the first good idea that comes to my mind.

17. When deciding on how to solve an open-ended problem (problem with no obvious solution or strategy), I weigh the efficiency of each strategy and compare them against each other.
18. When I make a plan to solve an open-ended problem (problem with no obvious solution or strategy), I am almost certain that I can make it work.
19. I try to predict the overall result of carrying out a particular course of actions.
20. When I try to think up possible solutions to an open-ended problem, I do not come up with very many alternatives.
21. Given enough time and effort, I believe I can solve most open-ended problems that confront me.
22. When faced with an open-ended problem that I don't understand immediately, I have confidence I can eventually solve it.
23. When I work on an open-ended problem (problem with no obvious solution or strategy), I feel like I do not know what to do to solve it.
24. When I work on an open-ended problem (problem with no obvious solution or strategy), I rush to find a solution.
25. I trust my ability to solve new and difficult problems.
26. I compare possible solutions to make a decision on which strategy to use.
27. When confronted with an open-ended problem (problem with no obvious solution or strategy), I do not usually think of other ways to solve it.
28. When I am confused by an open-ended problem (problem with no obvious solution or strategy), one of the first things I do is analyze what the problem requires and what I may need to do.
29. Sometimes I get confused, which keeps me from considering other ways of solving the problem.
30. After making a decision on how to solve an open-ended problem (problem with no obvious solution or strategy), the outcome I expected usually matches the actual outcome.
31. When confronted with a problem, I am not sure of whether I can solve it.
32. When I first encounter an open-ended problem (problem with no obvious solution or strategy), one of the first things I do is to determine relevant information needed to solve the problem.

REFERENCES

Association of Mathematics Teacher Educators. (2017). *Standards for Preparing Teachers of Mathematics.* Retrieved September 3, 2020, from amte.net/standards.

Avcu, S., & Avcu, R. (2010). Pre-service elementary mathematics teachers' use of stragegies in mathematical problem solving. *Procedia–Social and Behavioral Sciences, 9,* 1282–1286. https://doi.org/10.1016/j.sbspro.2010.12.321

Awofala, A. O. (2016). Examining preservice mathematics teachers' attitudes toward mathematics. *Nigerian Journal of Curriculum Studies, 23,* 292–300.

Asante, K. O. (2012). Secondary students' attitudes towards mathematics. *IFE PsychologIA: An International Journal, 20(*1), 121–133.

Aljaberi, N. M (2014). Pre-service elementary school teachers' level of mathematical thinking and their attitudes toward mathematics. *Journal of Education and Human Development, 3*(3), 181–195.

Bandura, A. (1986). *Social foundations of thought and action: A social cognition theory.* Prentice-Hall.

Bandura, A. (1994). Self-efficacy. In V. S. Ramachaudran (Ed.), *Encyclopedia of human behavior* (Vol. 4, pp. 71–81). Academic Press.

Bandura, A. (1997). Self-efficacy and health behavior. In A. Baum, C. McManus, J. Weinman, R. West, & S. Newman (Eds.), *Cambridge handbook of psychology, health and medicine* (pp. 160–162). Cambridge University Press.

Bal, A. P. (2014). The examination of representations used by classroom teacher candidates in solving mathematical problems. *Educational Sciences: Theory & Practice, 14*(6), 2349–2365.

Ball, D. L., & Bass, H. (2000). Interweaving content and pedagogy in teaching and learning to teach: Knowing and using mathematics. In J. Boaler (Ed.), *Multiple perspectives on the teaching and learning of mathematics* (pp. 83–104). Ablex.

Ball, D. L., & Forzani, F. M. (2011). Building a common core for learning to teach, and connecting professional learning to practice. *American Educator, 35*(2), 17–21, 38–39.

Bekdemir, M. (2010). The pre-service teachers' mathematics anxiety related to depth of negative experiences in mathematics classrooms while they were students. *Educational Studies in Mathematics, 75*(3), 311–328.

Briley, J. S. (2012). The relationships among mathematics teaching efficacy, mathematics self-efficacy, and mathematical beliefs for elementary pre-service teachers. *Issues in the Undergraduate Mathematics Preparation of School Teachers, 5,* 1–13.

Bruun, F. (2013). Elementary teachers' perspectives of mathematics problem solving strategies. *The Mathematics Educator, 23*(1), 45–59.

Budak, A., & Budak, İ. (2011). Assessing perceptions of pre-service teachers' teacher knowledge through portfolios. *Procedia–Social and Behavioral Sciences, 15,* 1376–1380.

Carrillo, J., & Cruz, J. (2016). Problem-posing and questioning: Two tools to help solve problems. In P. Felmer, E. Pehkonen, & J. Kilpatrick (Eds.), *Posing and solving mathematical problems: Advances and new perspectives* (pp. 23–36). Springer. https://doi.org/10.1007/978-3-319-28023-3

Cater, G. S., & Norwood, K. S. (1997). The relationship between teacher and students' belief about mathematics. *School Science and Mathematics, 97*(2), 62–67.

Cai, J. (2010). Commentary on problem solving heuristics, affect, and discrete mathematics: A representational discussion. In B. Sriraman & L. English (Eds.), *Theories of mathematics education* (pp. 251–258). Springer.

Chapman, O. (2013). Mathematical-task knowledge for teaching. *Journal of Mathematics Teacher Education, 16*(1), 1–6. https://doi.org/10.1007/s10857-013-9234-7

Council for the Accreditation of Educator Preparation. (2013). *Accreditation standards and evidence: Aspiration for educator preparation.* http://caepnet.org/standards/archive-standards/2013-itp

Education Northwest. (2011). *A guide for assessing mathematical understanding.*

Feikes, D., Schwingendorf, K., & Gregg, J. (2018). *Children's mathematical learning.* Purdue University Northwest.

Felmer, P., Pehkonen, E., & Kilpatrick, J. (Eds.). (2016). *Posing and solving mathematical problems,* Research in Mathematics Education Book Series. Springer-Verlag. https://doi.org/10.1007/978-3-319-28023-3_1

Gresham, G. (2008). Mathematics anxiety and mathematics teacher efficacy in elementary pre-service teachers. *Teaching Education, 19*(3), 171–184. https://doi.org/10.1080/10476210802250133

Gimbert, B., Bol, L., & Wallace, D. (2007). The influence of teacher preparation on student achievement and the application of national standards by teachers of mathematics in urban secondary schools. *Education and Urban Society, 40*(1), 91–117. https://doi.org/10.1177/0013124507303993

Guzman, G. M. (2018). Mathematical problem-solving strategies among teachers. *Journal on Efficiency and Responsibility in Education and Science, 11*(3), 53–64. https://doi.org/10.7160?eriesj.2018.110302

Harold-Blasius, R. (2017). Strategy keys as tools for problem solving. *Mathematics Teaching in the Middle School, 23*(3), 146–153. https://doi.org/10.5951/mathteacmiddscho.23.3.0146

Heppner, P. P., & Petersen, C. H. (1982). The development and implications of a personal problem-solving inventory. *Journal of Counsel Psychology, 29*(1), 66–75. https://doi.org/10.1037/0022-0167.29.1.66

Hiebert, J., Carpenter, T. P., Fennema, E., Fuson, K., Wearne, D., Murray, H., Oliver, A., & Human, P. (1997). *Making sense: Teaching and learning mathematics with understanding*. Heinemann.

Jitendra, A. K., Sczesniak, E., & Deatline-Buchman, A. (2005). An exploratory validation of curriculum-based mathematical word problem solving tasks as indicators of mathematics proficiency for third graders. *School Psychology Review, 34*, 358–371. https://doi.org/10.1080/02796015.2005.12086291

Jovanovic, J., & King, S. S. (1998). Boys and girls in the performance-based science classroom: Who's doing the performing? *American Educational Research Journal, 35*, 477–496. shttps://doi.org/10.3102/00028312035003477

Kirtman, L. (2008). Pre-service teachers and mathematics: The impact of service-learning on teacher preparation. *School Science and Mathematics, 108*(3), 94–102. https://doi.org/10.1111/j.1949-8594.2008.tb17812.x

Kloosterman, P., & Stage, F. K. (1992). Measuring beliefs about mathematical problem solving. *School Science and Mathematics, 92*(3), 109–115. https://doi.org/10.1111/J.1949-8594.1992.TB12154.X

Koichu, B., Berman, A. and Moore, M. (2004). Promotion of heuristic literacy in a regular mathematics classroom. *For the Learning of Mathematics, 24*(1), 33–39.

Kolovou, A., Van Den Heuvel-Panhuizen, M., & Bakker, A. (2009). Non-routine problem solving tasks in primary school mathematics textbooks—A needle in a haystack. *Mediterranean Journal for Research in Mathematics Education, 8(2)*, 31–68.

Köğce, D., Yıldız, C., Aydın, M., & Altındağ, R., (2009). Examining elementary school students' attitudes towards mathematics in terms of some variables. *Procedia Social and Behavioral Sciences, 1*(1), 291–295.

Klein, M. (2004). The premise and promise of inquiry based mathematics in pre-service teacher education: A poststructuralist analysis. *Asia-Pacific Journal of Teacher Education, 32*(1), 35–47. https://doi.org/10.1080/1359866042000206008

Lawsha, M., & Hussain, W. (2011). Secondary students' attitude towards mathematics in a selected school of maldives. *International Journal of Humanities and Social Science, 1*(15), 277–281.

Lambdin, D. V. (2003). Benefits of teaching through problem solving. In F. K. Lester (Ed.), *Teaching mathematics through problem solving: Prekindergarten—Grade 6* (pp. 3–13). National Council of Teachers of Mathematics.

Lesh, R. E., & Doerr, H. M. (2003). *Beyond constructivism: Models and modeling perspectives on mathematics problem solving, learning, and teaching*. Lawrence.

Lesh, R., & Zawojewski, J. (2007). Problem solving and modeling. In F. K. Lester Jr. (Ed.), *Second handbook of research on mathematics teaching and learning* (pp. 763–804). Information Age Publishing.

Leonard, J., Brooks, W., Barnes-Johnson, J., & Berry III, R. Q. (2010). The nuances and complexities of teaching mathematics for cultural relevance and social justice. *Journal of Teacher Education, 61*(3), 261–270. https://doi.org/10.1177/0022487109359927

Ma, L. (1999). *Knowing and teaching elementary mathematics: Teacher's understanding of fundamental mathematics in China and the United States.* Lawrence Erlbaum.

Ma, L. (2020). *Knowing and teaching elementary elementary mathematics: Teachers' understanding of fundamental mathematics in China and the United States*. Routledge.

National Academies of Sciences, Engineering, and Medicine. (2020). *Changing expectations for the K–12 teacher workforce: Policies, preservice education, professional development, and the workplace.* The National Academies Press. https://doi.org/10.17226/25603.

National Council of Teachers of Mathematics. (1980). *An agenda for action.*

National Council of Teachers of Mathematics. (2000). *Principles and standards for school mathematics*.

National Council of Teachers of Mathematics. (2014). *Principles to action: Ensuring mathematical success for all.* https://www.academia.edu/43141094/Principles_to_Actions_Ensuring_Mathematical_Success_for_All

National Research Council. (2001). *Adding it up: Helping children learn mathematics.* The National Academies Press. https://doi.org/10.17226/9822.

Neale, D. C. (1969). The role of attitudes in learning mathematics. *The Arithmetic Teacher, 16*, 631–640.

Nuangchalerm, P. (2012). Enhancing pedagogical content knowledge in preservice science teachers. *Higher education studies, 2*(2), 66–71. https://doi.org/10.5539/hes.v2n2p66

Olanoff, D. E., Kimani, P. M., & Masingila, J. O. (2009). Learning to teach via problem solving and supporting preservice teachers in learning via problem solving. In , S. L. Swars, D. W. Stinson, & S. Lemons-Smith (Eds.), *Proceedings of the 31st annual meeting of the North American Chapter of the International Group for the Psychology of Mathematics Education* (Vol. 5, pp. 1299–1307). Georgia State University.

Pajares, F. (1996). Self-efficacy beliefs and mathematical problem-solving of gifted students. *Contemporary Educational Psychology, 21*(4), 325–344. https://doi.org/10.1006/ceps.1996.0025

Pendergast, D., Garvis, S., & Keogh, J. (2011). Pre-service student-teacher self-efficacy beliefs: An insight into the making of teachers. *Australian Journal of Teacher Education, 36*(12), 46–58. https://doi.org/10.14221/ajte.2011v36n12.6

Pólya,, G. (1945). *How to solve it.* Princeton University Press.

Pólya, G. (1985). *How to solve it* (2nd ed.). Princeton University Press.

Philipp, R. A., Ambrose, R., Lamb, L. L. C., Sowder, J. T., Schappelle, B. P., Sowder, L., & Chauvot, J. (2007). Effects of early field experiences on the mathematical content knowledge and beliefs of prospective elementary school teachers: An experimental study. *Journal for Research in Mathematics Education, 38*(5), 438–476. https://doi.org/10.2307/30034961

Ramirez, G., Chang, H., Maloney, E. A., Levine, S., & Beilock, S. L. (2016). On the relationship between math anxiety and math achievement in early elementary school: The role of problem solving strategies. *Journal of Experimental Child Psychology, 141*, 83–100. https://doi.org/10.1016/j.jecp.2015.07.014

Siegle, D., & McCoach, D. B. (2007). Increasing student mathematics self-efficacy through teacher training. *Journal of Advanced Academics, 18*(2), 278–312. https://doi.org/10.4219/jaa-2007-353

Schoenfeld, A. H. (1985). *Mathematical problem solving*. Academic Press.

Schoenfeld, A. H. (2013). Reflections on problem solving theory and practice. *The Mathematics Enthusiast, 10*(1), 9–34.

Schmoker, M. (2006). *Results now: How we can achieve unprecedented improvements in teaching and learning*. Association of School Curriculum Development.

Schommer-Aikins, M., Duell, O. K., & Hutter, R. (2005). Epistemological beliefs, mathematical problem-solving beliefs, and academic performance of middle school students. *The Elementary School Journal, 105*(3), 289–304. https://doi.org/101086/428745

Sriraman, B., & Lesh, R. (2006). Modeling conceptions revisited. *ZDM, 38,* 247–254. https://doi.org/10.1007/BF02652808

Starko, A. J., & Schack, G. D. (1989). Perceived need, teacher efficacy, and teaching strategies for the gifted and talented. *Gifted Child Quarterly, 33*(3), 118–122. https://doi.org/10.1177/001698628903300305

Swars, S., Hart, L. C., Smith, S. Z., Smith, M. E., & Tolar, T. (2007). A longitudinal study of elementary pre-service teachers' mathematics beliefs and content knowledge. *School Science and Mathematics, 107*(8), 325–335. https://doi.org/10.1111/j.1949-8594.2007.tb17797.x

Temel, S. (2014). The effects of problem-based learning on pre-service teachers' critical thinking dispositions and perceptions of problem-solving ability. *South African journal of education, 34*(1), 1–20. http://www.sajournalofeducation.co.za

Telese, J. A. (2016). Improving elementary teachers' self-efficacy for mathematics teaching. *Teacher Education and Practice, 29*(4), 615–630.

Telese, J. A., Zhang, Z., & Diaz, M. (2020). The role of self-efficacy and other characteristics of elementary mathematics teachers: A model to predict student achievement. In P. Jenlink (Ed.), *A commitment to teaching: Toward more efficacious teacher preparation* (pp. 93–112). Rowman & Littlefield.

Van de Walle, J. A., Karp, K. S., & Bay-Williams, J. M. (2019). *Elementary and middle school mathematics: Teaching developmentally* (10th ed.). Pearson.

Wilson, M., & Cooney, T. J. (2002). Mathematics teacher change and development: The role of beliefs. In G. Leder, E. Pehkonen & G. Töerner (Eds.), *Beliefs: A hidden variable in mathematics education?* (pp. 127–147). Kluwer Academic.

Yilmaz, C., Altun, S. A., & Olkun, S. (2010). Factors affecting students' attitude towards math: ABC theory and its reflection on practice. *Procedia Social Science and Behavioral Sciences, 2,* 4502–4506.

CHAPTER 3

WHAT SHOULD WE TEACH IN ELEMENTARY MATHEMATICS METHODS COURSES?

The Argument for Core Pedagogical Content Knowledge in K–5 Mathematics Teacher Preparation

Courtney Glavich Mayakis and John A. Williams III

INTRODUCTION

Elementary education teachers graduating without in-depth content knowledge as well as the tools and instructional strategies to teach K–5 mathematics effectively has been discussed as an issue in teacher preparation today (National Research Council, 2003; Ryan & Williams, 2007; Stacey, et al., 2001). Even though many teacher education programs face challenges regarding various aspects of their programs, elementary teacher education presents unique challenges to teacher educators.

Since elementary teachers are tasked with teaching all content, teacher education programs, serving elementary preservice teachers (PTs), must ensure that they are gaining all the knowledge necessary to be successful

Mathematics as the Science of Patterns:
Making the Invisible Visible to Students Through Teaching, pp. 41–59

teachers. This requires teacher educators to carefully select what they teach PTs in their methods courses, often forcing teacher educators to "cut-out" vital aspects of pedagogical content knowledge. This can have detrimental effects on their teacher effectiveness; especially as it relates to mathematics.

However, Mewborn (2001) notes that various studies (i.e., Ball & Wilson, 1990, Fuller, 1997; Ma, 1999) that have employed qualitative data methods to understand the complexities of content and teaching practices, found that elementary education teachers who were strong in mathematics content were not always strong in conceptual understanding, dismissing the idea that strong content background equates to a strong conceptual background. Fennema and Franke (1992) concluded "that when a teacher has a conceptual understanding of mathematics, it influences classroom instruction in a positive way" (p. 151). Mewborn (2001) asserts that various research studies (i.e., Ball & Wilson, 1990; Fuller, 1997; Ma, 1999) have suggested that even though elementary education teachers may have an understanding of the algorithms and procedures that are utilized in their curriculum, they do not have a conceptual understanding of the mathematics, and that "their knowledge tends to be compartmentalized and fragmented and, therefore, not easily transferable from one domain to another" (Mewborn, 2001).

However, if a strong background in content knowledge is not enough, how are we developing PTs into effective mathematics teachers who are not only able to teach their content, but able to conceptually understand and instruct their mathematics curriculum? One thought is that elementary teacher programs are not providing enough coursework surrounding mathematics in their preservice programs. The American Council on Education (1990) has suggested that more mathematics coursework can be related to increased student learning. However, Monk (1994) discovered that this was true only up to a total of five courses, in which student achievement becomes diminished.

Wilkins (2008) suggests that "being good" in mathematics or simply taking more courses in mathematics may not provide PTs with the sufficient background to effectively teach mathematics instruction in their classroom, and argues that there may be other factors which influence this. These factors, or the effective teaching practices that elementary education programs should focus on as an organizational framework, can be examined through the lens of Shulman's (1986) three types of teaching knowledge.

Schulman's Three Types of Teaching Knowledge

According to Shulman (1986) there are three types of knowledge that teachers must understand in order to be successful which are; (1) *Content*

Knowledge, which refers to not only the facts and knowledge of a specific discipline but also the structures of this discipline; (2) *Pedagogical Content Knowledge,* which pertains to the content knowledge and strategies which are specific to teaching that particular content; and (3) *Curricular Content Knowledge,* which refers to the knowledge of a program or set of instructional materials that is required for a specific content for a certain grade level.

For the purposes of analyzing contemporary mathematical standards and preservice learning, we attest that the significance of content and pedagogical knowledge in K–5 mathematics is vital to the effectiveness of teachers in the classroom (Belge, 1979; Cochran et al., 1993; Kahanet al., 2003). Hill et al. (2008) contextualize Schulman's three types of teaching knowledge through the lens of effective mathematics instruction.

The authors propose that not only do mathematics teachers possess content subject matter, but also have different types of knowledge that are useful for effective mathematics instruction. For preservice teachers, it is critical for them to understand how content within mathematics connects, as well have the fundamental capabilities to instruct based on the needs of their students (pedagogical knowledge), all the while utilizing relevant materials (curricular content knowledge).

For example, the instruction of a lesson on multiplication, would require a teacher to understand where his or her students are conceptually, any background knowledge they have on multiplication (utilizing previous lessons on addition). The successful application of the lesson would be highly dependent on the teachers' use of aids or tools (e.g., the number line, grouping, etc.) as the curriculum content.

Hill et al. (2008), define this other knowledge as KCS or "content knowledge intertwined with knowledge of how students think about, know or learn this particular content" (p. 375). This type of knowledge concentrates on how students learn through a certain topic, and what misconceptions that student may typically have when learning a certain concept. The authors argue that KCS is an extension of Shulman's (1986) PCK or

> an understanding of what makes the learning of specific topics easy or difficult: the conceptions and preconceptions that students of different ages and backgrounds bring with them to the learning of those most frequently taught topics and lessons. (p. 9)

Schulman supports Hill et al. (2008) by highlighting that understanding student thinking is a key component in pedagogical content knowledge. Ball and Bass (2000) define pedagogical content knowledge as the intertwining of content and pedagogical knowledge, and argue that this is a key component of teacher knowledge is often disconnected throughout instruction and practice.

Another component to his knowledge is, specialized content knowledge (SCK), or the understanding of the mathematical operations necessary to reach the correct answer for a problem (Ball et al., 2008; Hill et al., 2008). With these types of teacher knowledge being crucial in teacher development, are the standards and frameworks used to guide teacher preparation addressing these adequately?

Guiding Standards for Elementary Mathematics Teacher Education

Stakeholders in teacher quality and accreditation of teacher education programs have developed mathematical frameworks and general standards for K–12 teacher education to ensure that PTs are learning critical teaching practices. These mathematical frameworks, for initial teacher licensure, are based on research and best practices for mathematics education (Association of Mathematics Teacher Educators, 2013; Council for the Accreditation of Educator Preparation, 2015; National Council of Teachers of Mathematics, 2012).

These organizations have created reformed mathematical standards for K--12 students that focus on student-centered instruction, shifting from the once teacher-oriented instructional environment (Wilson et al., 2013). With most of these organizations focusing on K–12 classroom mathematical standards, some are now starting to develop or refine K–12 teacher education standards for mathematics.

Mathematical frameworks for initial teacher preparation were not readily discussed or developed until the 1980s when the National Council for Accreditation of Teacher Education (NCATE), in conjunction with the National Council of Teachers of Mathematics (NCTM), developed the first set of scholarly mathematics teacher preparation standards (NCTM, 1998). According to these program standards, institutions of higher education should respond to these standards if their program includes the mathematics teacher preparation of future mathematics teachers' in

> (1) middle school, (2) junior high school mathematics, and/or (3) senior high mathematics education. A program review for elementary education must be prepared if the program offers an area of concentration in mathematics or an emphasis in mathematics in the preparation of K–4 teachers. (NCTM, 1998, p. 2)

These first set of program/teacher educator standards created by NCTM in 1998, has evolved from a standard checklist, to a set of indicators and standards from several entities that embody rigorous themes related to mathematics instruction and content in teacher education programs (AMTE, 2013; CAEP, 2015; NCTM, 1998).

Several of these entities (AMTE, 2013; CAEP, 2015; NCTM, 2012) capture mathematical frameworks situated in secondary education, as well as some in elementary education mathematics programs. There are two sets of mathematical frameworks or mathematical recommendations that will be synthesized in this review of the literature; The Council for the Accreditation of Educator Preparation (CAEP) and the Association of Mathematics Teacher Educators (AMTE).

The Council for the Accreditation of Educator Preparation's (CAEP) K–5 Elementary Teacher Standards is the leading set of K–5 teacher educator standards used currently (general standards). The Association of Mathematics Teacher Educators (AMTE), recently released program standards for teacher preparation for elementary mathematics that allows for a specific outline of essential mathematical standards for elementary PTs.

CAEP K–5 Elementary Teacher Standards and AMTE Upper Elementary and Early Childhood Standards

In an effort to strengthen teacher education programs and restructure the accreditation process, CAEP set forth a new set of standards for K–5 elementary teachers. The standards are intended to trim down the accreditation process and allow programs to focus more on improving their graduates' impact on student achievement (Heafner et al., 2014). The CAEP K–5 elementary teacher standards consist of five cohesive standards supported by key elements to guide teacher educators on how to implement these standards in their instruction, however, for this literature review, we will be focusing only four.

The Association of Mathematics Teacher Educators (AMTE) is another professional organization that addresses mathematical standards with a focus on recommendations for teacher candidates. AMTE has recently released a comprehensive set of standards used to guide the instruction of all P–12 teacher education programs, as well as minimum requirements regarding coursework.

AMTE suggests that 12-semester hours of mathematical education of teachers as well as a minimum of 6 weeks of statistical education of teachers should be included in P-5 mathematics teacher education (AMTE, 2017). AMTE groups these mathematical standards, for the primary grades, into two sections, early childhood mathematics, and upper elementary preparation. Standards for both CAEP and AMTE's two sections displayed in Figure 3.1.

Figure 3.1

AMTE and CAEP Standards for K–5 Mathematics Teachers

Upper Elementary Mathematics Standards (AMTE)	Details	Early Childhood Mathematical Standards (AMTE)	Details	CAEP K-5 Elementary Teacher Standards	Details
1. Mathematics Concepts, and Connections to Mathematical Practices	Understanding of foundational mathematic concepts, and the ability to connect those concepts to mathematical practices in Pre-K-2	1. Mathematics Concepts, and Connections to Mathematical Practices	Should possess robust knowledge of mathematical and statistical concepts. Can read, discuss and analyze curriculum, assessment, and standards, as well as students' mathematical procedures	1. Understanding and Addressing Each Child's Developmental and Learning Needs	Must have ability to understand child growth and development, individual differences with families, cultures and communities, to produce high quality learning experiences.
2. Pedagogical Knowledge and Teaching Practices	Develop pedagogical knowledge and practices to cultivate students mathematical proficiency towards conceptual understanding, procedural fluency, problem-solving, and facilitation with mathematical processes.	2. Pedagogical Knowledge and Teaching Practices	Have mathematical foundations of pedagogical knowledge, effective and equitable mathematic teaching practices, and positive and productive dispositions toward teaching mathematics to support students' understanding, and reasoning.	2. Understanding and Applying Content and Curricular Knowledge for Teaching	Demonstrate and apply understandings of major concepts, skills, and practices as they interpret standards across literacy, mathematics, science, and social studies for K-6 grades.
3. Students as Learners of Mathematics	Effective use of tools, physical models, and mathematical representations to build student understanding.	3. Students as Learners of Mathematics	Have foundational understanding of students mathematical knowledge, skills, and dispositions, and are committed to deepening students' knowledge of mathematics.	3. Assessing, Planning, and Engaging Learners for Instruction	The planning and adaptation of instruction sequences to promote a full range of competencies for every learner base on knowledge of each child, educational goals, instructional practices, disciplinary knowledge, and student assessment through social norms.
		4. Social Context of Mathematics and Teaching and Learning	The understanding and the application of social, historical, and institutional contexts of mathematics.	4. Supporting Each Child's Learning Using Effective Instruction	Use of informal decision about instruction, guided by knowledge of children and assessment of students' learning that results in a variety of effective learning

Note: Adapted from Association of Mathematics Teacher Educations [AMTE]. (2017). *Standards for preparing teachers of mathematics* and Council for the Accreditation of Educator Preparation (CAEP). (2012). *Draft CAEP K–5 elementary teacher standards.*

In examining the CAEP mathematics curricular standards addressed in Standard 2, it is evident that one key element is focused on the major mathematical domains in elementary teaching.

> Candidates demonstrate and apply understandings of major mathematics concepts, algorithms, procedures, applications and mathematical practices in varied contexts, and connections within and among mathematical domains (Number and Operations, Operations and Algebraic Thinking, Measurement and Data [both Statistics and Probability] and Geometry) as presented in the rationale for the CAEP Mathematics Content for Elementary (K–5) Teachers. (CAEP, 2012, p. 2)

However, this is the only portion of the CAEP standards in which mathematical practices and content are explicitly named. However, AMTE provides elementary teacher educators with the specific, in-depth standards (e.g., AMTE addressing in detail the pedagogical knowledge needed in Standard 2) that the CAEP standards do not provide teacher educators with (e.g., CAEP addresses major mathematical concepts that intersect into other content areas for Standard 2).

The overarching standards within CAEP leave teacher educators with ambiguous paths to follow, as they try to guide preservice teachers towards effective mathematical practices for student learning. Even though ATME separates the content and pedagogical knowledge needed for PTs to be successful in early childhood and upper elementary mathematics, they give content specific examples of how these two knowledges are intertwined.

In the standards for early childhood and upper elementary education, AMTE details the content that is essential to preparing elementary PTs. Even though they provide specific content examples, they do not discuss how these units of content can be connected to PCK under a broader conceptual idea. They argue that these content exemplars are "specific but not exhaustive examples" (AMTE, 2017, p. 49).

AMTE (2017) details a cohesive learning trajectory for subitizing in the early childhood mathematics, but does not give an example developmental learning trajectory (LTs) for upper elementary, and does not discuss how LTs can be utilized in these grades. AMTE describes tasks in Part 2 (programmatic features) that teacher educators could utilize with PTs in regards to PCK in both early childhood and upper elementary mathematics.

However, these tasks are not interconnected under overarching frameworks, and do not link to other singular units of key pedagogical content knowledge. Even though these tasks could be helpful to teacher educators, it does not detail how these particular tasks could translate into a teacher educators' classroom (i.e., assignments that would be given, assessment, teaching techniques, etc.), or what successful instruction if these interconnected knowledges (content and pedagogy) would entail.

Guiding Frameworks for Elementary Mathematics Teacher Education

One might raise a question as to whether these standards created for teacher educators are enough to support the learning of PTs mathematical conceptual knowledge in elementary education. In order for PTs to understand and deliver effective mathematics instruction (outputs), effective mathematical knowledge to teach (inputs) must be provided.

Even though CAEP and AMTE offer standards and recommendations and characteristics that can guide the instruction of PTs, instructional frameworks, such as learning trajectories and cognitive guided instruction, also guide the way that teacher educators instruct their PTs. These frameworks are focused on best practices and student thinking and learning, and help create a context in which to deliver mathematical standards.

Learning trajectories. Timmermans (2010) explains that "the most significant aspect of learning lies not in the outcomes of learning, but in the process of learning" (p. 3). However, to fully capture the efficiency and rigor of the learning process in respective curricula, one must understand the key concepts in their curricular specific learning processes or these "trajectories of learning" (p. 3).

Perkins (2007) explains that LT's are "especially pivotal to a stage-like advance in understanding a discipline" (p. 36). Research shows that knowing and understanding mathematical learning trajectories can assist teachers in better sequencing activities and instructional tasks to better fit the needs of their students (Clements & Sarama, 2004; Daro et al., 2011; Sztajn et al., 2012).

Simon (1995) stated that a "hypothetical learning trajectory" included "the learning goal, the learning activities, and the thinking and learning in which the students might engage" (p. 133). Clements and Samara (2004) mirror this definition by arguing that a learning trajectory consists of three components, the actual learning goal or objective, the developmental progressions of thinking and learning, and sequence of instructional tasks to accomplish this objective. Confrey et al. (2009) states

> a researcher-conjectured, empirically-supported description of the ordered network of constructs a student encounters through instruction (i.e., activities, tasks, tools, forms of interaction and methods of evaluation), in order to move from informal ideas, through successive refinements of representation, articulation, and reflection, towards increasingly complex concepts over time. (p. 347)

However, Battista (2004) argues that learning trajectories are sections or layers of conceptual thinking in which students pass from pre-instructional concepts, progressing through various other, inherently more sophisticated concepts, concluding their understanding at a recognized mathematical concept.

However, the inherent goal of learning trajectories is to study developmental progressions in conjunction with a sequence of instructional tasks to create an optimal learning environment in order for students to absorb an instructional goal. Learning trajectories or LTs, once developed, will eventually lead students to transform their conceptual thinking of mathematical concepts. They produce concepts from trajectories of learning that can transform a way a student thinks about a particular concept.

Cognitive guided instruction. According to Carpenter and Fennema (1991), cognitive guided instruction (CGI) as a framework for student thinking, is the idea that "teachers use knowledge from cognitive science to make their own instructional decisions" (p. 10). Vacc and Bright (1999) assert that, through this type of instruction, teachers use their research-based knowledge regarding student learning, to plan and lead the design of their instruction, and therefore seek to understand individual students' thinking.

Teachers who utilize CGI principles believe eliciting and understanding student's thinking is a key component in their instructional planning. They lead their students in problem solving and exploring of concepts, utilize this problem solving in determining their students' conceptions by listening and responding with questions until their thinking becomes clear and then proceed to make instructional mathematical decisions that fit the needs of their students (Fennema et al., 1989).

With theoretical underpinnings in Vygotsky (1987) CGI, "provides a basis for teachers to interpret, transform, and reframe their informal or spontaneous knowledge about students' mathematical thinking" (p. 5). Due to these theoretical underpinnings, a large factor of how well a teacher implements the CGI principles rely on their beliefs about mathematics instruction and learning (Fennema et al., 1998). The CGI framework enables elementary education mathematics teachers to establish and understand students' thinking, while creating instructional strategies that foster student's conceptual understanding on the basis of their current conceptual thinking.

These frameworks and standards guide teacher educators' instruction of PTs. Even though these standards and frameworks focus on a variety of aspects of teaching and learning, such as content, instructional strategies, student thinking and learning, and students' progressions of thinking, they lack a cohesive interweaving of content and pedagogical knowledge, and therefore fail to focus on pedagogical content knowledge. However, the significance of content and pedagogy, and the conceptual power of bridging these knowledges to form pedagogical content knowledge, are of great importance to effective teaching.

Significance of Content Knowledge

When discussing K–5 mathematics student learning, it is often heard that students would learn more math if their teachers knew more math (Kahan et al., 2003). In a study conducted on teachers' knowledge of mathematics, Belge (1979) concludes that the importance of profoundly understanding and knowing one's content knowledge is frequently dismissed in elementary teacher education. We know that the mathematics content knowledge that PTs possess is vital to their teaching success (Marks, 1990), and has been consistently deemed important in the field of teaching.

Since the No Child Left Behind Act, signed into office in 2002, the idea of "highly qualified teachers" has been in the forefront. Highly qualified teachers is defined in this act as a teacher that has, (1) earned a bachelor's degree, (2) obtained full state certification or licensure, and (3) proven that they know each subject they teach (No Child Left Behind, 2002).

What encompasses proving they know each subject they teach can depend on the measure set forth by each state. For example, in the state of North Carolina, K–5 PTs acquiring a teaching license must pass A Core Academic Skills for Educators for writing, reading and mathematics (Educational Testing Services, 2016). The mathematics test consists of a total of 56 questions, with number and quantity consisting of 30% of the test, algebra and functions also consisting of 30% of the tests, geometry and statistics consisting of 20% of the test with probability consisting of 20% of the test as well.

These tests consist of content that teachers should know in order to be successful in teaching in their field. Due to the rigorous content on many of these mathematics tests, some states are finding it difficult for teacher candidates to pass their required tests. States such as South Dakota and Michigan have written locally about the problem that their aspiring teachers are facing (Anderson, 2015; French, 2013).

As Hill et al. (2005) explain, teacher knowledge has typically been measured by coursework or degrees attained. However, there has been a recent development in the literature that has looked closer at teacher's knowledge in terms of the "ability to understand and use subject-matter knowledge to carry out the tasks of teaching" (p. 372), originally described by Shulman (1986).

However, are these important concepts (content knowledge and pedagogical knowledge) being integrated in initial licensure standards and/or instruction? For example, the CAEP standard that singularly addresses mathematics concepts, algorithms, procedures and practices, and connections throughout mathematics domains, is limited in scope. It does not address the type of advanced content knowledge that is required to increase learners' understanding of the different connections between mathematical domains. Specifically looking at content, there have been numerous studies that look closely at different content specific aspects of each domain that are vital to student learning in K–5 mathematics.

For example, the development of fractional concepts is readily discussed in the literature, as well as an emphasis on measurement, patterns, number sense, and the development of algebraic thinking, just to name a few of the underlying key conceptual practices in early mathematics (Cai et al. 2005; Clements & Stephan, 2004; Empson, 1999; Mulligan & Mitchelmore, 2009; Reys, 1991; Van de Walle et al., 2007). If these specific mathematical concepts, that change the way of teacher thinking and practice, and consequently student learning, are not being addressed interconnectedly in the current standards, how are we ensuring that these are the concepts that are being taught in universities?

Significance of Pedagogical Knowledge

It is vital that teachers are able to become experts in all three areas of knowledge as it relates to their curriculum. However, there has been a special interest on pedagogical knowledge. The word pedagogy was derived from the Greek word "peadagogue" which means a keeper, rather than a teacher (Hall, 1905). It has recently evolved to mean the art, or science of teaching. In terms of teacher knowledge, many argue whether prospective teachers should major in education or if they should major in a discipline (Ball & Bass, 2000).

Dewey (1904/1964) classified good teachers as teachers who generate "genuine intellectual activity" in their students. He also believed that the "methods" on how to achieve this were inevitably tied to disciplines. "Despite these prescient ideas, teacher education throughout the 20th century has consistently been structured across a persistent divide between subject matter and pedagogy" (Ball & Bass, 2000, p. 242). Ball and Bass (2000) explain how this divide is often discussed within debates of equity and reaching all students.

As Hooks (2014) explains, engaged pedagogy is a way to make the learning process more effective for our students. We should not only look at teaching as a way to share information but a way to foster our students' curiosity and understanding. Given the research and conceptualization of what effective pedagogy can do for students in the classroom, it is inevitable that effective teaching and learning of pedagogy by prospective teachers could support equitable outcomes in school.

Previous literature has shown that pedagogical techniques in a K–5 mathematics setting, or any setting for that matter, would result in effective learning and understanding for students. Teachers that are able to explain more than one strategy for solving mathematical problems will have a better chance of reaching students who would otherwise may not understand the one method that a teacher without a strong pedagogical background might teach.

According to Ball et al. (2008) "Although the term pedagogical content knowledge is widely used, its potential has been thinly developed" (p. 389). It has also lacked empirical evidence and definition that has limited its usefulness in the teaching world. Defining pedagogical knowledge and what it is, is scarcely studied empirically.

Besides Shulman's original work, there have been little efforts to create well-specified accounts of what students know in this area (Cochran et al., 1993). Without empirical proof of what teachers need to know in order to be successful, these claims have a minimal impact on furthering the field of education (Foss & Kleinsasser, 1996).

The Case for Intertwining Content and Pedagogy: Strengthening Pedagogical Content Knowledge

Spearheaded by Schulman's (1986) three types of teacher knowledge, a growing body of literature has "begun to conceptualize teachers' knowledge for teaching differently, arguing that teacher effects on student achievement are driven by teachers' ability to understand and use subject-matter knowledge to carry out the tasks of teaching" (Hill et al., 2005, p. 372), instead of conceptualizing teachers' knowledge using measures outputs such as courses taken, grades in courses, or scores on skills test (Kleickmann et al., 2013). In other words, this growing argument specifies that student achievement is specifically directed to *what* teachers know and *how* they use it to instruct.

Therefore, we argue that content (what) and pedagogy (how) should be focused on equitability to ensure that there is an undeniable focus on both as equivalent entities to the importance of primary mathematics teaching. The current standards in usage for K–5 mathematics teacher educators lack a core focus on pedagogical practices and content in breadth and depth of core ideas and understandings related to K–5 mathematics teaching. Historically, content has been the main focus of teacher education courses and programs. In order to increase teacher effectiveness in the K–5 mathematics classroom both important entities of teaching, pedagogy and content, need to be addressed.

Since Shulman (1986) stressed the importance of teachers being immersed in pedagogy and content, teacher education programs have sought to infuse these within their programs, and mathematical accrediting organizations have strived to align their standards to meet this challenge. Still, despite studies linking teacher pedagogical effectiveness in math to student achievement (Campbell et al., 2014; Clotfelter et al., 2010), current teacher mathematical standards lack frameworks that address pedagogy and content *together*.

Even though AMTE has recently written standards that embed pedagogy into content, and describes some of these notions as well as gives example tasks for preservice teachers, they are not exhaustive or extensive interconnected ideas of pedagogy and content, that link to overarching key mathematical concepts. Linking overarching key ideas to singular pedagogical and content knowledge can allow preservice teachers to understand mathematics in a conceptual way in primary mathematics. In a study of international teacher preparation programs and mathematical standards, Tatto et al. (2010) discovered a low emphasis on instructing primary mathematics teachers on pedagogy in the United States.

Among the countries that place a low emphasis on teacher education in mathematics, their study determined that teachers found themselves resembling general education teachers, rather than mathematics teachers (Tatto et al., 2010). The lack of integrated mathematical content and pedagogical knowledge within the standards forces teachers to develop on their own, what may or may not be effective solutions to their own deficiency (Daro et al., 2011). This effort tends to have teachers relying on ineffective methods, instead of developing more efficient methods of instruction.

A CASE FOR CORE PEDAGOGICAL CONCEPT KNOWLEDGE IN K-5 MATHEMATICS TEACHER EDUCATION

In conclusion, it is evident throughout the literature that good teacher education can be directly related to effective teaching. However, it is also apparent that elementary mathematics' teachers graduate without the essential conceptual knowledge to effectively instruct their students (Ball & Bass, 2000; Campbell et al., 2014; Clotfelter et al., 2010; Mewborn, 2001).

Many relate this lack of conceptual knowledge and teacher readiness to the inherent separation of content and pedagogy (Ball & Bass, 2000; Campbell et al., 2014; Clotfelter et al., 2010). As Shulman (1986) discusses, pedagogical content knowledge is a key component in the delivering of instruction and contextualizing content.

Yet, the interweaving of content and pedagogy, or pedagogical content knowledge is lacking in current teacher education programs and practicums (Ball & Bass, 2000), even though this connection could strengthen knowledge and instruction (Dewey, 1904/1964). Therefore, we seek to understand how these special forms of knowledge can be taught through the interconnection of singular units of pedagogical content knowledge that link to an overarching mathematical concept that can span domains and grade levels. We argue that the development of threshold concepts in elementary mathematics teacher education can assist in this connection.

THRESHOLD CONCEPTS

A threshold concept is a key idea that can open a new and previously unreachable way of thinking about a something, once a student has internalized this idea (Meyer & Land, 2003).

> Students who have not yet internalized a threshold concept have little option but to attempt to learn new ideas in a more fragmented fashion. On acquiring a threshold concept, a student is able to transform their use of the ideas of a subject because they are now able to integrate them in their thinking. (Meyer & Land, 2003, p. 53)

Threshold concepts have distinct characteristics that define their importance. They are considered *transformative*, or in other words, once a threshold concept is internalized is has the power to change the ability that an individual sees a subject matter.

To put this in terms of elementary mathematics, once a preservice teacher understands the idea of decomposing and composing numbers utilizing the base ten numeration system, they would be able to explain to their students what "borrowing" conceptually means in the traditional algorithm of subtraction instead of explaining only the procedure. It will change the way that teacher candidates view and understand numbers and traditional algorithms for various operations.

Threshold concepts are also considered *irreversible*. Once an individual understands a threshold concept, it is not easily forgotten. Using the above example, a teacher candidate would find it difficult to forget that numbers can be decomposed and composed by groups represented in our base ten system, and would find it hard to not explain this concept when teaching multi-digit subtraction, addition, and so forth.

Threshold concepts are characterized by being *integrative* or are defined by exposing interrelated concepts, that before seem disconnected. For example, in early elementary mathematics, subitizing is considered a foundational skill for elementary students to grasp. However, if teacher candidates have developed a foundational background in a threshold concept related to the base ten numeration system, they would understand that subitizing can be connected to more advanced skills such as skip counting or accessing numbers in groups other than one, which would lead to the ability to "see numbers" and utilize this skill to break apart multidigit numbers to make 10s.

However, threshold concepts can be *troublesome* to learners. For example, teacher candidates who have only learned or understood the procedural method of multidigit subtraction, addition, and so forth. could find the conceptual understanding of decomposing numbers to solve these more complex problems difficult. Many teacher candidates entering teacher education programs do not have a deep understanding of the conceptual knowledge behind procedures, which could require teacher educators to have more conversations around these concepts.

Threshold concepts are also *bounded* which means they can only be utilized in that specific discipline. For example, the understanding of base 10 and its characteristics could be considered a threshold concept in elementary mathematics. Even though base 10 spans grade levels, it is considered unique to the mathematics discipline.

The gap discussed previously in the article between pedagogy and content underscores a key area where threshold concepts can inform practice related to teachers' ability to apply and effectively teach specific content, as well as the instructional tasks that are used to develop teacher candidates' understanding of this content. Teacher candidates' ineffectiveness to deliver inadequate instruction to students is partly due to their inability to evolve past their current procedural understanding of concepts.

CONCLUSIONS

Using threshold concepts in mathematics would serve to change the conceptualization process of preservice teachers, ultimately fostering effective mathematics instruction once they are classroom teachers. By identifying these threshold concepts in K–5 mathematics education, teacher educators can have frameworks to utilize in order to ensure that preservice teachers are learning the content and pedagogical knowledge that is vital to successful mathematics instruction for K–5 students.

Threshold concepts would also inform what is the *most* important to teach and which concepts have the largest impact on teacher candidates' learning and therefore their subsequent teaching. Elementary education programs are unique in that they are tasked with educating their teacher candidates' on how to instruct and become an expert in four content areas.

Therefore, universities only having time to instruct one or two elementary mathematics content. If we are unable to teach all that an elementary teacher needs to know in their preservice mathematics courses, then what do we focus on?

The insertion of SCK would be useful as a guide for teacher educators to discern key interconnected concepts that are essential for preservice teachers to understand, and how these overarching concepts (PCK, CK, and SCK) can relate to smaller units of curriculum and specific teaching techniques. However, we urge the further research and discussion regarding *what* should be included with describing these frameworks, ultimately leading to the discovery and use of threshold concepts in K-5 mathematics teacher education.

REFERENCES

American Council on Education. (1990). *To touch the future. Transforming the way teachers are taught.*

Anderson, P. (2015). *Future teachers can't pass a math test.* Retrieved October 10, 2017, from http://www.argusleader.com/story/news/education/2015/03/25/future-teachers-pass-math-test/70459508/.

Association of Mathematics Teacher Educations [AMTE]. (2017). *Standards for preparing teachers of mathematics.* Retrieved March 25, 2017, from https://amte.net/sites/default/files/SPTM.pdf.

Association of Mathematics Teacher Educators (AMTE). (2013). *Review of the CAEP standards draft.* Meredith College, AMTE. https://amte.net

Ball, D. L., & Bass, H. (2000). Interweaving content and pedagogy in teaching and learning to teach: Knowing and using mathematics. In J. Boaler (Ed.), *Multiple perspectives on the teaching and learning of mathematics* (pp. 83–104). Ablex.

Ball, D. L., Thames, M. H., & Phelps, G. (2008). Content knowledge for teaching: What makes it special? *Journal of Teacher Education, 59*(5), 389–407. https://doi.org/10.1177/0022487108324554

Ball, D. L., & Wilson, S. M. (1990). *Knowing the subject and learning to teach it: Examining assumptions about becoming a mathematics teacher.* National Center for Research on Teacher Education, Michigan State University.

Battista, M. (2004). Applying cognition-based assessment to elementary school students' development of understanding of area and volume measurement. *Mathematical Thinking and Learning, 6*(2), 185–204. https://doi.org/10.1207/s15327833mtl0602_6

Belge, E. G. (1979). *Critical variable in mathematics education: Findings from a survey of the literature.* Mathematical Association of America

Cai, J., Lew, H. C., Morris, A., Moyer, J. C., Ng, S. F., & Schmittau, J. (2005). The development of students'' algebraic thinking in earlier grades. *Zentralblatt für Didaktikder Mathematik, 37*(1), 5–15. https://doi.org/10.1007/BF02655892

Campbell, P. F., Nishio, M., Smith, T. M., Clark, L. M., Conant, D. L., Rust, A. H., & Choi, Y. (2014). The relationship between teachers' mathematical content and pedagogical knowledge, teachers' perceptions, and student achievement. *Journal for Research in Mathematics Education, 45*(4), 419–459.

Carpenter, T. P., & Fennema, E. (1991). *Research and cognitively guided instruction.* In E. Fennema, T. P. Carpenter, & S. J. Lamon (Eds.), *Integrating research on teaching and learning mathematics* (pp. 1–16). State University of New York Press.

Clements, D. H., & Sarama, J. (2004). Learning trajectories in mathematics education. *Mathematical Thinking and Learning, 6*(2), 81–89. https://doi.org/10.1207/s15327833mtl0602_1

Clements, D. H., & Stephan, M. (2004). Measurement in pre-K to Grade 2 mathematics. In D. H. Clements, J., Sarama, & A. M. DiBiase (Eds.), *Engaging young children in mathematics: Standards for early childhood mathematics education* (pp. 299–317). Lawrence Erlbaum Associates.

Clotfelter, C. T., Ladd, H. F., & Vigdor, J. L. (2010). Teacher credentials and student achievement in high school a cross-subject analysis with student fixed effects. *Journal of Human Resources, 45*(3), 655–681. https://doi.org/10.1353/jhr.2010.0023

Cochran, K. F., DeRuiter, J. A., & King, R. A. (1993). Pedagogical content knowing: An integrative model for teacher preparation. *Journal of Teacher Education, 44*, 263–272. https://doi.org/10.1177/0022487193044004004

Confrey, J., Maloney, A., Nguyen, K., Mojica, G., & Myers, M. (2009). Equipartitioning/splitting as a foundation of rational number reasoning using learning trajectories. In M. Tzekaki, M. Kaldrimidou, & H. Sakonidis (Eds.), *Proceedings of the 33rd Conference of the International Group for the Psychology of*

Mathematics Education—in search for theories in mathematics education: PME 33. Thessaloniki, Greece, July 19–24, 2009.

Council for the Accreditation of Educator Preparation (CAEP). (2012). *Draft CAEP K–5 elementary teacher standards*. CAEP. www.caepnet.org.

Daro, P., Mosher, F. A., & Corcoran, T. B. (2011). *Learning trajectories in mathematics: A foundation for standards, curriculum, assessment, and instruction*. http://repository.upenn.edu/

Dewey, J. (1964). *The educational situation* (No. 3). University of Chicago Press. (Original work published 1904)

Educational Testing Services. (2016). *North Carolina test requirements*. www.ets.org

Empson, S. B. (1999). Equal sharing and shared meaning: The development of fraction concepts in a first-grade classroom. *Cognition and Instruction, 17*(3), 283–342. https://doi.org/10.1207/S1532690XCI1703_3

Fennema, E., Carpenter, T. P., Franke, M. L., Levi, L., Jacobs, V. R., & Empson, S. B. (1998). A longitudinal study of learning to use children's thinking in mathematics instruction. *Journal for Research in Mathematics Education, 29*(1), 3–20. https://doi.org/10.3102/0013189X027005006

Fennema, E., & Franke, M. L. (1992). Teachers' knowledge and its impact. In D. A. Grouws (Ed.), *Handbook of research on mathematics teaching and learning* (pp. 147–164). Macmillan.

Fennema, E., Carpenter, T. P., & Peterson, P. L. (1989). Teachers' decision making and cognitively guided instruction: A new paradigm for curriculum development. In N. F. Ellerton & M. A. Clements (Eds.), *School mathematics: The challenge to change* (pp. 174–187). Deakin University Press.

Foss, D. H., & Kleinsasser, R. C. (1996). Preservice elementary teachers' views of pedagogical and mathematical content knowledge. *Teaching and Teacher Education, 12*(4), 429–442. https://doi.org/10.1016/0742-051X(95)00049-P

French, R. (2013). Passing Michigan's teacher tests is a breeze—and learning suffers. Retrieved October 10, 2017, from http://www.bridgemi.com/special-report/passing-michigans-teacher-tests-breeze-and-learning-suffers

Fuller, R. A. (1997). Elementary teachers' pedagogical content knowledge of mathematics. *Mid-Western Educational Researcher, 10*(2), 9–16.

Hall, G. S. (1905). What is pedagogy? *The Pedagogical Seminary, 12*(4), 375–383. https://doi.org/10.1080/08919402.1905.10534667

Heafner, T., McIntyre, E., & Spooner, M. (2014). The CAEP standards and research on educator preparation programs: Linking clinical partnerships with program impact. *Peabody Journal of Education, 89*(4), 516–532. https://doi.org/10.1080/0161956X.2014.938998

Hill, H. C., Ball, D. L., & Schilling, S. G. (2008). Unpacking pedagogical content knowledge: Conceptualizing and measuring teachers' topic-specific knowledge of students. *Journal for Research in Mathematics Education, 39*(4), 372-400.

Hill, H. C., Rowan, B., & Ball, D. L. (2005). Effects of teachers' mathematical knowledge for teaching on student achievement. *American Educational Research Journal, 42*(2), 371–406. https://doi.org/10.3102/00028312042002371

Hooks, B. (2014). *Teaching to transgress*. Routledge.

Kahan, J. A., Cooper, D. A., & Bethea, K. A. (2003). The role of mathematics teachers' content knowledge in their teaching: A framework for research applied to a study of student teachers. *Journal of Mathematics Teacher Education, 6*(3), 223–252. https://doi.org/10.1023/A:1025175812582

Kleickmann, T., Richter, D., Kunter, M., Elsner, J., Besser, M., Krauss, S., & Baumert, J. (2013). Teachers' content knowledge and pedagogical content knowledge the role of structural differences in teacher education. *Journal of Teacher Education, 64*(1), 90–106. https://doi.org/10.1177/0022487112460398

Ma, L. (1999). *Knowing and teaching elementary mathematics: Teachers' understanding of fundamental mathematics in China and the United States*. Lawrence Erlbaum Associates.

Marks, R. (1990). Pedagogical content knowledge: From a mathematical case to a modified conception. *Journal of Teacher Education, 41*(3), 3–11. https://doi.org/10.1177/002248719004100302

Meyer, J., & Land, R. (2003). Threshold concepts and troublesome knowledge—Linkages to ways of thinking and practising (pp. 53–64). In C. Rust (Ed.), *Improving student learning—Ten years on*. Oxford Center for Staff and Learning Development.

Mewborn, D. (2001). Teachers content knowledge, teacher education, and their effects on the preparation of elementary teachers in the United States. *Mathematics Teacher Education and Development, 3,* 28–36. https://doi.org/10.1177/0022487112460398

Monk, D. H. (1994). Subject area preparation of secondary mathematics and science teachers and student achievement. *Economics of Education Review, 13*(2), 125–145. University of Edinburgh.

Mulligan, J., & Mitchelmore, M. (2009). Awareness of pattern and structure in early mathematical development. *Mathematics Education Research Journal, 21*(2), 33–49. https://doi.org/10.1007/BF03217544

National Council of Teachers of Mathematics (NCTM). (1998). *NCATE (NCTM) Program standards*.

National Council of Teachers of Mathematics (NCTM). (2012). *NCTM CAEP standards—Elementary mathematics specialist (advanced preparation)*.

National Research Council. (2003). *Adding it up: Helping children learn mathematics*. National Academies Press.

No Child Left Behind (NCLB) Act of 2001, Pub. L. No. 107–110, § 115, Stat. 1425 (2002).

Perkins, D. (2007). Theories of difficulty. In N. Entwistle & P. Tomlinson (Eds.), Student learning and university teaching (pp. 31–48). British Psychological Society.

Reys, B. J. (1991). *Developing number sense. Curriculum and evaluation standards for school mathematics addenda series, Grades 5–8*. National Council of Teachers of Mathematics.

Ryan, J., & Williams, J. (2007). *Children's mathematics 4–15: Learning from errors and misconceptions*. McGraw-Hill.

Shulman, L. S. (1986). Those who understand: Knowledge growth in teaching. *Educational Researcher, 15*(2), 4–14. https://doi.org/10.3102/0013189X015002004

Simon, M. A. (1995). Reconstructing mathematics pedagogy from a constructivist perspective. *Journal for Research in Mathematics Education, 26*(2), 114–145. https://doi.org/10.2307/749205

Stacey, K., Helme, S., Steinle, V., Baturo, A., Irwin, K., & Bana, J. (2001). Preservice teachers' knowledge of difficulties in decimal numeration. *Journal of Mathematics Teacher Education, 4*(3), 205–225. https://doi.org/10.1023/A:1011463205491

Sztajn, P., Confrey, J., Wilson, P. H., & Edgington, C. (2012). Learning trajectory based instruction toward a theory of teaching. *Educational Researcher, 41*(5), 147–156. https://doi.org/10.3102/0013189X12442801

Tatto, M. T., Lerman, S., & Novotna, J. (2010). The organization of the mathematics preparation and development of teachers: a report from the ICMI Study 15. *Math Teacher Education, 13*(4), 313–324. https://doi.org/10.1007/s10857-009-9139-7

Timmermans, J. A. (2010) Changing our minds. In Meyer, J. H., Land, R., & Baillie, C. (Eds.). *Threshold concepts and transformational learning*. Sense.

Vacc, N. N., & Bright, G. W. (1999). Elementary preservice teachers' changing beliefs and instructional use of children's mathematical thinking. *Journal for Research in Mathematics Education, 30*(1), 89–110. https://doi.org/10.2307/749631

Van de Walle, J. A., Karp, K. S., & Williams, J. M. B. (2007). *Elementary and middle school mathematics. Teaching Development*. Pearson.

Vygotsky, L. S. (1987). Thought and word In R. W. Rieber & A. S. Carton (Eds.), *The collected works of L. S. Vygotsky* (Vol. 1, pp. 243–285). Plenum.

Wilkins, J. L. (2008). The relationship among elementary teachers' content knowledge, attitudes, beliefs, and practices. *Journal of Mathematics Teacher Education, 11*(2), 139–164. https://doi.org/10.1007/s10857-007-9068-2

Wilson, P. H., Mojica, G. F., & Confrey, J. (2013). Learning trajectories in teacher education: Supporting teachers' understandings of students' mathematical thinking. *The Journal of Mathematical Behavior, 32*(2), 103–121. https://doi.org/10.1016/j.jmathb.2012.12.003

CHAPTER 4

EFFECTIVENESS OF DIAGNOSTIC INTERVIEWS

Preservice Teachers' Patterns of Conceptions in Student Understanding of Number and Operation

Diana L. Moss, Lisa L. Poling, and Tracy Goodson-Espy

INTRODUCTION

Interviewing, a balance between questioning and observing, in a mathematics classroom becomes a heuristic alternative to traditional forms of assessment, allowing individuals to grow in their own mathematical understanding, as well as serving the needs of their students (Huinker, 1993; National Council of Teachers of Mathematics [NCTM], 1991). Various types of interviews have a long and rich history in mathematics education.

Interviews may be used as a mechanism for documenting students' current levels of understanding (clinical or diagnostic interviews), or as a means of exploring current knowledge and building upon it (teaching experiments) (Steffe & Thompson, 2000). Interviews are also deemed to be useful for helping the interviewer (preservice teachers [PSTs]) learn how to focus on an aspect of the teaching and learning process, such as

Mathematics as the Science of Patterns:
Making the Invisible Visible to Students Through Teaching, pp. 61–73

questioning (Moyer & Milewicz, 2002). Posner and Gertzog (1982) state, regardless of initial intent, the primary function of any interview protocol is to identify and acknowledge the depth, breadth, and related connections among content that an individual retains.

Crespo (2000) in her analysis of PSTs' abilities to conduct an interview states that, "for prospective teachers, the idea of listening to students is not obvious" (p. 156). The beliefs held by PSTs about what it means to do mathematics may remain conventional, assuming that mathematical proficiency is only established through identifying a correct response using a set of rules and procedures (Moyer & Moody, 1998); therefore "preservice teachers need to be engaged in experiences that encourage them to reflect on their growing knowledge of how students think mathematically" (Moyer & Moody, 1998, p. 613).

As PSTs engage in interview processes and analyze the mathematical activity of students, a realization that mathematics is multifaceted might begin to emerge (Schorr et al., 2013). As PSTs encourage and listen to student ideas regarding problem solving in mathematics, they may also construct a deeper understanding of mathematics, modifying their beliefs about what it means to teach and learn mathematics (Fennema & Carpenter, 1996; Moyer & Milewicz, 2002).

PSTs should learn the purposes for, and benefits of, various types of interviews because teachers are often called upon to conduct diagnostic interviews for assessment purposes, as well as use interviews to extend a child's understanding of a mathematical concept. The following section presents the theoretical framework for the study.

THEORETICAL FRAMEWORK

This study describes the results from a diagnostic interview assignment (DIA) completed within a mathematics content and pedagogy class. The PSTs in this study must take a series of three mathematics content and pedagogy courses. The second course emphasizes the content strand of number and operation, and within the field experience, helps PSTs focus on the mathematical activity and thinking of individual children and small groups.

PSTs complete a cycle of instructional tasks aimed at helping them deepen their content knowledge of subtraction, while also demonstrating a process for learning any content topic. In the course, PSTs are exposed to a variety of activities and literature pertaining to subtraction and foundational topics. The conceptual literature from early childhood mathematics (subitizing, counting, basic operations, children's thinking, approaches to algorithms, etc.) that supports the DIA include:

1. Subitizing and unitizing (Clements, 1999; Clements & Sarama, 2014; van de Walle et al., 2012)
2. Addition and subtraction problem structures as described within cognitively guided instruction (CGI) (Carpenter et al., 1999)
3. Children's counting strategies as described within CGI
4. Children's invented algorithms (van de Walle et al., 2012)
5. Children's responses to formal algorithms (Bass, 2003; Philipp & Cabral, 2005; van de Walle et al., 2012)
6. Procedural versus conceptual understanding (Baroody, 2003; Hiebert, 1986; Rittle-Johnson & Koedinger, 2009)

The DIA occurs during the field experience and is focused on subtraction. The purpose of the DIA is to: (1) to allow PSTs to observe how subtraction constructs that they have studied in the methods class unfold in a classroom context; (2) to illustrate how to conduct a diagnostic interview as opposed to a teaching interview; (3) to encourage them to describe student understandings of subtraction, based on their understanding of mathematics education literature; and (4) to allow them to see differences in how children respond to the same instruction concerning subtraction.

Below, we describe the DIA and the methodology employed in the study. This study aimed to explore the following questions:

1. How do PSTs interpret patterns of conception in students' mathematical thinking based on diagnostic interviews with second and third-grade students?
2. How do PSTs reflect on, and discuss, future implications for teaching based on their analyses of student thinking?

METHODOLOGY

A broader study was conducted during the spring semester at a four-year public university in the Southeastern United States. The participants ($n = 22$) for this study were enrolled in a mathematics methods course that met twice a week for 2 hours and 20 minutes twice every week for 11 weeks. Data collection occurred in the last 5 weeks of the semester during the practicum for the course.

This report describes a snapshot of data from this broader study that considers the effect of content/methods professors' instructional decisions on the outcome for PSTs in terms of the development of their mathematical understanding, as well as their abilities to examine and characterize children's mathematical activity. For this research report, three PSTs were chosen from the full diagnostic interview task for a more in-depth analysis based on the following factors: their interview was conducted with second or third-grade students, and the PSTs' high achievement in the class.

All participants are female, Caucasian, and between 20 and 21 years of age. The qualitative data for this study came from the PSTs' written work on the DIA required to complete the course and written transcripts from the interviews.

Design of the Diagnostic Interview

The purpose of the DIA is to assess PSTs' ability to follow students' thinking as they work through mathematics activities. The DIA provided PSTs with an interview template on subtraction and required them to conduct an interview with three students. They took notes, collected student work, audio recorded and transcribed the interviews, and wrote a reflection about student thinking and the interview process.

The DIA was assigned during the last four weeks of the semester when PSTs were in elementary classrooms to observe teachers and students (see Table 4.1). In the reflection, PSTs were asked to provide background information on the students and school setting, analyses of students' responses to the questions, implications for teaching, and their impressions of the interview process. An abbreviated set of sample instructions is shown in Table 4.1.

Analysis of Data

To explore the research questions of how PSTs interpreted students' mathematical thinking and how they reflected on, and discussed, future implications for teaching based on their analyses of student thinking, we read and examined the DIA for constructs, themes, and patterns. The researchers then coded the themes using a constant comparative method (Strauss, 1987).

Three project researchers read and coded the DIA papers independently using the following questions: (1) Did the PSTs specify the mathematics content and discuss it correctly?; (2) Did the PSTs focus on the child's answer and/or the child's mathematical thinking?; (3) Did the PSTs address and explain whether the child focused on a procedure to complete the problems or did the child rely on a more conceptual approach?; and (4) Based on the previous questions, what implications for teaching did the PSTs suggest?

Following analyses, the researchers were able to recognize two main themes within the PSTs' responses. In the following section, we provide

Table 4.1

Diagnostic Interview Assignment

Diagnostic Interview Assignment
Procedures: Your objective is to record what the student currently knows about the concept. Remember that the purpose of this interview is to assess what the child knows and can do rather than report what he/she cannot do. Show the child one problem at a time. Ask the child, "*Can you solve this problem?*" Make brief notes about what the child does and says. If the child asks a question or makes a comment, do not provide the answer at this point. Instead, ask the child, "*How could you figure that out?*" Do not tell the child how to solve the problem. If the child is not successful with the initial tasks, modify the tasks until you find the point at which the child begins to experience some success. Throughout the interview ask questions such as, "*How did you do that?*" **Let the student do the work and do not teach.** Follow-up on the student's work or comments with questions, such as, "*What were you thinking?*" or "*Why did you do it that way?*" • With the help of your cooperating teacher, identify three students to interview. Based on academic performance in math, select a high, medium, and low performing student. • Be sure to collect all of the work that students do during the interview. Soon after the interview, write down any impressions you had during the interview. Audio tape each interview and transcribe loosely. • Analyze your notes/transcripts and write a paper about each student's mathematical understanding related to the interview problems and reflection about the implications for teaching. Use pseudonyms for the students. You will need to use examples of student work or conversation as evidence.
Analysis of Students Responses • Background information—Describes the students and the setting. • What mathematics does each student understand? • How would you characterize his/her understanding? Be sure to think about conceptual versus procedural understanding and use this language in your characterization about each student's understanding. • Which CCSS standards are listed for subtraction and/or place value for the grade level of the students you interviewed? How does this influence your interpretation of each student's reasoning? • What differences did you notice between the three students? What tools did the students use?
Implications for Teaching • Based on how you characterized your students' understanding, what recommendations would you make to build on these students' mathematical understanding? • What remediation or enrichment would be appropriate for each student given the scope of your interview? • How do these interviews influence the way you think about teaching this content?

(Table continued on next page)

Table 4.1 (Continued)

Diagnostic Interview Assignment

Diagnostic Interview Assignment
Interview Plan 2nd & 3rd Grades (Subtraction)
Materials Needed: Problems, index card, base-ten manipulatives, Unifix cubes, paper and pencil.
Problems (Tasks) Selected for the Student Interview and the Rationale for the Problems Chosen: These problems were selected to help us learn about the child's understanding of whole number subtraction. Tell him/her you are going to ask some math questions and they can use any of the materials to solve the problems.
Pam has 35 stickers. John gave her 17 more stickers. How many stickers does Pam have now? • Pam had 48 stickers. She bought some more stickers. Now she has 62 stickers. How many stickers did she buy? • Pam has 52 stickers. John has 25 stickers. How many more stickers does Pam have than John? • 44 – 17 = [Write the problem horizontally on a large index card] • 44 – 17 = [Write the problem vertically on the other side of the index card] • Can you think of another way to solve the last problem? [If the student did not use any of the materials, ask them if they could solve it using one of the manipulatives provided.] • Can you make up a subtraction word problem for me to solve?

the DIA analyses of three PSTs as they discuss the mathematical activities of their interviewee and their impressions of the child's mathematical understanding.

RESULTS

The results are presented in this section, with the interpretation of student's mathematical activity and implications for future teachers discussed, based on the analysis of data drawn from the full diagnostic interview task. A subset of three selected preservice teachers' interviews provides the basis for the results.

Hilary's Interpretation of the Student's Mathematical Activity

Hilary (PST) interviewed Linda, an average second-grade student who indicated that she does not enjoy mathematics. Hilary described the content that Linda understood from a positive, encouraging perspective, but she explains Linda's problem-solving activities by acknowledging what Linda does not do as opposed to what she can do in terms of the mathematics. Hilary was overly concerned with the child applying keyword strategies to solve the problem. For instance, she explained the content that the student was able to do as:

> This student understands adding and subtracting when borrowing and/or carrying is not involved. She knows that there needs to be borrowing and carrying in these problems, but gets confused on when to do which. She is great at interpreting word problems, and determining which operation to use *based on the context clues.*

The next excerpt also refers to Linda, but pertains to how she is thinking as opposed to her content knowledge.

> I would characterize her understanding as wavering. She possesses procedural knowledge, acknowledging that there are steps that she needs to follow in each of the operations given a certain scenario. However, she often gets these steps confused, because she does not understand the "why" behind them. This being said, her conceptual understanding is greatly lacking.

Hilary explains what Linda can do procedurally but when Linda is unable to do a procedure, then Hilary interprets this as a lack of conceptual understanding. Hilary seems to believe that if the child is able to perform a procedure, such as the subtraction algorithm, correctly, then this implies that the child possesses a conceptual understanding of the concept.

Hilary's Suggestions for Implications for Future Teaching

Hilary explains that Linda has a procedural understanding of subtraction, but lacks a conceptual understanding. Hilary explains that the reasons she thinks that the student does not have a conceptual understanding is:

> If it (conceptual knowledge) were stronger, she would have no issue with problems that involve carrying and borrowing.... In her second problem, "62–48," she would need to borrow from the 6 to be able to subtract 8 from 2. However, she just pretended you could reverse the minuend and subtrahend, and subtract 2 from 8, so she could put down 6 and would not have to borrow.

According to Hilary, Linda's lack of understanding of carrying and borrowing in subtraction is a gap in conceptual understanding. However, carrying and borrowing is a procedure of the standard algorithm, and thus, this explanation implies that Hilary's analysis that the student lacks conceptual understanding is insubstantial, when, in fact, the diagnostic interview suggests that the student does not understand subtraction procedurally. In the last part of the DIA, Hilary recommends that the teacher use base ten blocks to help Linda understand the concepts of borrowing and carrying. Hilary concludes:

> Upon seeing and understanding why the steps occur, this should in turn strengthen her procedural understanding of these operations because she will be able to look at the problem and understand why she needs to perform whichever step.

Hilary does not mention conceptual understanding when writing about the implications of the diagnostic interview for teaching this student. Her recommendation suggests that Linda needs to understand certain steps of performing subtraction and this indicates that Hilary is focused on the standard algorithm of subtraction and, therefore, a procedural understanding of subtraction. Based on this diagnostic interview, it is unclear if Hilary knows what conceptual understanding is and how to teach it.

Elayne's Interpretation of the Student's Mathematical Activity

Elayne (PST) identified that James (a student identified as average achieving in mathematics), maintained a number of competencies in his mathematical understanding. She stated that James knew the standard algorithm, used multiple subtraction strategies, can comprehend and analyze word problems, has place value knowledge, can properly use manipulatives and understands the concepts of borrowing and regrouping. Elayne recorded that one example of his depth of understanding was when attending to the second question of the interview.

The question states "Pam has 48 stickers. She bought some more stickers. Now she has 62 stickers. How many stickers did she buy?" To solve this problem James wrote 48 + __ = 62. He understood that 62 – 48 is the same as adding something to 48 in order to get 62. In each of the problems, James explained how he solved it with a large amount of base ten knowledge. He knew the importance of place value (adding/subtracting the ones place first, then the tens.)

Using addition to solve a subtraction problem is a valuable strategy; however, Elayne, in this example, only focused on the operations, failing to recognize the overall demand of the problem and what it was representing. An appropriate question, to check for conceptual understanding, might have been "What does the __ (answer) represent?" A question such as this may have given Elayne more insight into what she has identified as his skills.

Elayne later stated that she is "impressed with James' level of conceptual and perceptual (procedural) understanding of borrowing and regrouping with addition and subtraction. Although in question 3 (Pam has 52 stickers. John has 25 stickers. How many more stickers does Pam have than John?") she states:

> To solve this, James wrote 52 + 25 = ____. He stated that 2 + 5 is seven, and then wrote 37 as his answer. When asked to explain, he said "I got 37 because 2 plus 5 is seven but then I realized that three plus two is five." Essentially, he was adding from the bottom up. In his verbal explanation through, he said that 5 plus 2 is 7. If he truly were adding from the bottom up, he would have said "7 plus 5 is 12" and then would have figured out that 3 must be added to 2 in order to get 5. Regardless, he was able to arrive at the correct answer, but based on his verbal explanation.

What is concerning about this exchange is that Elayne did not recognize that James, in fact, had not obtained a correct answer and that his strategy was not consistent. He stated he adds 2 + 5 to get 7, but then changes his techniques and solves __ + 2 = 5. If adding up was James' choice of strategy then his second equation should have been 5 + 2 = 7 and his answer would have been 77. In this problem, James shows neither conceptual nor procedural understanding.

Elayne's Suggestions for Implications for Future Teaching

Elayne identified James as having "deeper level of understanding of the concepts" and he displayed a "higher level of both conceptual and procedural understanding" based on his diagnostic interview. She also states that James has a "basic understanding of subtraction and place value concepts." Her recommendations for attending to these gaps was to focus on improving his mathematical fluency through the use of flashcards, teach him more subtraction strategies and let him invent his own strategies.

Elayne, after completing the diagnostic interview with James struggles to clearly identify what is known and unknown. Elayne's evaluation is contradictory and fails to identify some larger gaps that exist in the student's knowledge, which also indicate gaps in Elayne's knowledge regarding subtraction. Throughout her analysis, Elayne focused on keywords related to subtraction and mathematical understanding, but failed to accurately identify student understanding and failed to suggest productive teaching implications.

Erin's Interpretation of the Student's Mathematical Activity

Erin was able to provide a detailed description of the mathematics her highest achieving student, Marlon, was able to perform and discuss. She noted that he was able to complete the first two problems from the interview quickly, in his head, and did not show any written work. She then prompted Marlon to show her how he completed the second problem, using Unifix cubes as a tool. Problem 2 states, "Pam had 14 stickers. She bought some more stickers. Now she has 19 stickers. How many stickers did she buy?"

In response, Marlon counted out 14 blocks on one color and then added blocks of a second color until he reached 19. Erin described this activity as, changing a subtraction problem into an addition problem (14 + __ = 10 rather than 19-14 = ___). Erin did not describe this problem structure in terms of missing-addend, but she did seem to understand that there were multiple ways to look at it. Erin also noted that Marlon was using a counting-up strategy to solve the problem.

For the third problem, "Pam has 18 stickers. John has 7 stickers. How many more stickers does Pam have then John?", Erin explained that Marlon was initially unable to solve the problem. He was unwilling to draw a picture, but was willing to use the Unifix cubes. Erin noted that Marlon created one stick with 18 cubes and a second stick with 7 cubes and moved them in the air. She then explained that because Marlon did not line the sticks up, he was not able to count from 7 to 18 successfully.

As Erin asked him to lay the blocks on the table and recount, he was able to find the correct answer. Erin did not distinguish between problem 1 and problem 2 as being different in structure and she did not refer to problem 3 as a comparison problem. Erin explained that Marlon stated at the conclusion of problem 3 that, it would have been easier to have one stick of 18 and then take way seven cubes.

Concerning the part of the interview where the student was asked to solve the same problem in vertical rather than horizontal form (13 – 5 = ___), Erin stated that Marlon stated that he saw the problem differently when it was stated in vertical notation. She did not explain what Marlon meant by this, however. Marlon was able to come up with his own method to solve the problem. 10 – 5 = 5 and then 3 + 5 = 8.

While Erin described this method, she did not characterize this as an invented strategy, despite being exposed to these ideas in her mathematics methods course. She did use this problem solution to argue that Marlon understands place value.

Erin's Suggestions for Implications for Future Teaching

While Erin was able to provide a detailed description of Marlon's solution activity, she did not seem to have a good handle herself on the underlying subtraction problem structures and how they impact children's problem-solving activity, or how advanced Marlon's counting strategies were for a second-grade student. While Erin described Marlon's understanding in positive terms when she described his responses to the interview questions, and moved to discussing implications for teaching, her analysis moved toward a deficit model.

She did not appreciate the advanced nature of Marlon's thinking and was overly concerned that he was not already fluent with his math facts. She, therefore, recommended working with subtraction using flash cards and completing activities involving multiplication facts. She was disturbed that he and his classmates used their fingers as manipulatives. Erin did, however, use the fact that Marlon was able to create his own subtraction word problem as evidence that he was developing a conceptual understanding of subtraction.

It is curious that she pointed to this as evidence despite the fact that she did not use different subtraction problem structures herself in describing Marlon's activity. Erin also stated that because Marlon was able to solve some of the problems in his head that this was evidence that he was moving from concrete to abstract thinking. She did not explain what she meant by these terms, however.

IMPLICATIONS

The three PSTs reported their observations of students' mathematical activity; however, these observations were not always correct. In particular, Hilary assumes that her interviewee does not have a conceptual knowledge of subtraction based on the mathematical procedures that the child is able or not able to perform. Elayne fails to notice that the mathematics that her interviewee performs is not correct, indicating that Elayne does may not have complete conceptual understanding of the content of subtraction.

Lastly, Erin stated that her interviewee had a concrete understanding of subtraction that moved to an abstract understanding because he could do subtraction in his head, when this could be simple memorization. The suggestions that the PSTs made for implications for future teaching were also flawed because they were based on incorrect evaluations of the mathematical content observed.

PSTs inability to comprehend the full extent of conceptual understanding regarding subtraction was recognized through the diagnostic interview analysis. Evidences from this study indicate that PSTs need additional opportunities to critically view and respond to children's mathematical thinking.

CONCLUSIONS

The purpose of this study was to determine the effectiveness of a diagnostic interview assignment for helping preservice teachers. The study sought to determine how well course readings and instructional activities prepared PSTs to analyze the results of the diagnostic interviews. The authors performed a backward analysis of PSTs' diagnostic interview papers to determine the patterns of conceptions within the assessment of how PSTs interpreted students' mathematical thinking and how they reflected on, and discussed, future implications for teaching based on their analyses of student thinking.

As demonstrated through this research, not only did the PSTs discover gaps in children's understanding, we as mathematics educators also gained another vantage point from which to gain access to the PST's conceptual understanding of subtraction. Recognizing the gaps in PSTs ability to critical diagnose student thinking, provides awareness into how we engage and evaluate classroom discussions grounded in student thinking.

This study supports that conducting diagnostic interviews provides PSTs with the opportunity to learn about mathematical thinking (Schorr et al., 2013) and to practice listening to and making sense of students' ideas (Crespo, 2000). We offer themes that characterize the mathematical activity that PSTs notice and their suggestions for implications for future teaching based how they reflected on their analyses of student thinking.

REFERENCES

Baroody, A. J. (2003). *The development of adaptive expertise and flexibility: The integration of conceptual and procedural knowledge*. Erlbaum.

Bass, H. (2003). Computational fluency, algorithms, and mathematical proficiency: One mathematician's perspective. *Teaching Children Mathematics, 2*, 322–327. https://doi.org/10.5951/TCM.9.6.0322

Carpenter, T. P., Fennema, E., Franke, M. L., Levi, L., & Empson, S. B. (1999). *Children's mathematics: Cognitively guided instruction*. Heinemann.

Clements, D. (1999). Subitizing: What is it? Why teach it? *Teaching Children Mathematics, 5*, 400–405. https://doi.org/10.5951/TCM.5.7.0400

Clements, D., & Sarama, J. (2014). *Learning and teaching early math: The learning trajectories approach* (2nd ed.). Routledge.

Crespo, S. (2000). Seeing more than right and wrong answers: Prospective teachers' interpretations of students' mathematical work. *Journal of Mathematics Teacher Education, 3*, 155–181. https://doi.org/10.1023/A:1009999016764

Fennema, E., & Carpenter, T. P. (1996). A longitudinal study of learning to use children's thinking in mathematics instruction. *Journal for Research in Mathematics Education, 27*(4), 403–434. https://doi.org/10.2307/749875

Hiebert, J. (1986). *Conceptual and procedural knowledge: The case of mathematics.* Erlbaum.

Huinker, D. M. (1993). Interview: A window to students' conceptual knowledge of the operations. In N. L. Webb (Ed.), *Assessment in the mathematics classroom* (pp. 80–86). NCTM.

Moyer P. S., & Milewicz, E. (2002). Learning to questions: Categories of questioning used by preservice teachers during diagnostic mathematics interviews. *Journal of Mathematics Teacher Education, 5*, 293–315. https://doi.org/10.1023/A:1021251912775

Moyer, P. S., & Moody, V. R. (1998). *Shifting beliefs: Preservice teachers' reflection on assessing students' mathematical ideas* [Paper presentation] The 1998 North American Chapter of the International Group for the Psychology of Mathematics Education (PME-NA) Annual Meeting, Raleigh, NC.

National Council of Teachers of Mathematics. (1991). *Professional standards for teaching mathematics.*

Philipp, R., & Cabral, C. (2005). *Gretchen clip three. IMAP: Integrating mathematics and pedagogy to illustrate children's reasoning* [Computer software]. San Diego University Foundation. Retrieved September 28, 2010, from http://www.sci.sdsu.edu/CRMSE/IMAP/main.html

Posner, G. J., & Gertzog, W.A. (1982). The clinical interview and the measurement of conceptual change. *Science Education, 66*(2), 195–209. https://doi.org/10.1002/sce.3730660206

Rittle-Johnson, B., & Koedinger, K. R. (2009). Iterating between lessons concepts and procedures can improve mathematics knowledge. *British Journal of Educational Psychology, 79*, 483–500. https://doi.org/10.1348/000709908X398106

Schorr, R. Y., Arias, C. C, & Warner, L. B. (2013). *Prospective teachers' analysis of children's errors using clinical interviews* [Paper presentation]. The 2013 North American Chapter of the International Group for the Psychology of Mathematics Education (PME-NA) Annual Meeting, Chicago, IL.

Steffe, L. P., & Thompson, P. W. (2000). Teaching experiment methodology: Underlying principles and essential elements. In R. Lesh & A. E. Kelly (Eds.), *Research design in mathematics and science education* (pp. 267–307). Erlbaum.

Strauss, A. (1987). *Qualitative analysis for social scientists.* Cambridge University.

van de Walle, J. A., Karp, K. S., Bay-Williams, J. M., & Wray, J. (2012). *Elementary and middle school mathematics: Teaching developmentally* (8th ed.). Pearson.

CHAPTER 5

USING VISUAL REPRESENTATIONS TO EXPLORE PROPORTIONAL REASONING WITH FUTURE TEACHERS

Sheri Johnson and Rui Kang

INTRODUCTION

To qualify as a language, a communication system must have vocabulary, grammar, syntax, and a group of people who understand and use it. From a linguistic perspective, Wakefield (2000) recognizes mathematics as a language that has abstractions, uniform and consistent symbols and rules, and linear and serial expressions. Developing the language of mathematics is an essential component of mathematical proficiency (Riccomini et al., 2015). The CCSS Mathematical Practices (National Governors Association Center for Best Practices & Council of Chief State School Officers, 2010) explicitly address communication as a separate practice and a new point of emphasis. Constructing viable arguments and critiquing the reasoning of others (CCSS Mathematical Practice 3) encourages explanation in addition to correct answers, and the importance of language is weaved among other practices as well. For example, tending to precision (CCSS Mathematical Practice 6) includes

Mathematics as the Science of Patterns:
Making the Invisible Visible to Students Through Teaching, pp. 75–97

using mathematically appropriate vocabulary, which can be especially challenging for both students and teachers. From geometry, angles can be complementary or supplementary, from algebra, functions have a domain and range, and from probability and statistics, distributions can be normal and uniform. These are terms that have other meanings outside a mathematics classroom and understanding these terms in a mathematical context requires new connections to be formed.

The shared attributes between language and mathematics such as abstractions (e.g., symbols and rules) and translations (e.g., among symbols, images, and meanings) makes us realize the challenges facing the mathematics learner, as learning mathematics is sometimes not much different than learning a foreign language (Wakefield, 2000). Traditional, text-centered instructional settings often create barriers between the learner and mathematics (Riccomini et al., 2015). The notion of mathematics as a language calls for mathematics instruction to incorporate activities with rich contextual and visual cues.

In most classrooms in the United States, "mathematics is presented as an almost entirely numeric and symbolic subject, with a multitude of missed opportunities to develop visual understanding" (Boaler et al., 2016, p.1). Even when visual representations are incorporated, they are limited to graphs and tables that are often used to accompany a procedural explanation. Limiting visual representations to graphs and tables does not provide our students the rich opportunity to develop conceptual understanding, and they are often left with only procedural fluency, lacking the knowledge of how and when to apply their learned algorithms to real-world situations. As an alternative, our study is grounded in a problem-solving approach and extends visual representations to include drawings and diagrams, where participants designed their own models. We begin by presenting our theoretical perspectives and reviewing pertinent literature, then describe our study, and finally identify implications for education.

THEORETICAL PERSPECTIVES

Proportional reasoning is developed over time and extends beyond numerical fluency and reasoning with fractions. Working with a variety of situations and different visual representations deepens learning and allows students to recall when and how to determine if a proportional relationship exists. We value the creation and use of representations, recognize the challenge of quantitative reasoning, and rely on former research of proportional reasoning.

Visual Representations and Reasoning Across Representations

Visual representations are often thought to include "the big three," which are equations, graphs, and tables because these are what is typically used in mathematics education (Kaput, 1992). These are powerful tools that can reveal patterns and relationships, however there are additional meaningful visual representations including drawings, pictures, number lines, flow charts, diagrams, and more. Arcavi (2003) noted that visuals can "group together clusters of information that can be apprehended at once" (p. 218) and "visualization at the service of problem solving, may also play a central role to inspire a whole solution, beyond the merely procedural" (p. 224). Whether created or provided, visual representations allow exploration and aid learners in making sense of problems. Connecting visual representations such as symbols and drawings with their meaning and generating expressions and equations that define relationships are essential skills for successful mathematics learning.

Research-based guiding principles promote making connections across mathematical representations (National Council of Teachers of Mathematics, 2014). Dreher and Kuntze (2015) noted that reasoning across multiple representations is necessary for learners to develop an appropriate concept image of mathematical objects such as fractions. However, the translation between a mathematical representation and the object it stands for is not necessarily obvious and self-explanatory. When a student reasons with a new representation, connections have to be made among representations, and this challenging cognitive process can create obstacles (Dreher & Kuntze, 2015). These obstacles are an impetus for learning and in order for learners to benefit from multiple representations, they must be actively engaged in creating connections among various representations. In a mathematics classroom, this process not only allows students to think abstractly and quantitatively to advance their knowledge but also helps them develop their mathematical language, and this in turn deepens their understanding.

For teachers to successfully facilitate students in reasoning across multiple representations, they must see the function of these representations in a mathematics classroom as not only making instruction more engaging and diverse, but also playing an important role of helping students deepen their conceptual understanding (Dreher & Kuntze, 2015). Teachers must also develop the necessary pedagogical content knowledge related to the effective use of various representations (Shulman, 1986). For example, different content domains such as ratios and proportions often call for content-specific representations (Dreher & Kuntze, 2015). Teachers need to develop knowledge about how to select the most useful and powerful representations in specific instructional situations. They also need to be able to anticipate and interpret the types of reasoning likely involved with various representations, the difficulties students tend to encounter with each of the representations, why these difficulties might occur, and strategies for scaffolding students learning. To achieve these ends, we argue that mathematics educators must engage future teachers themselves in mathematics activities that require them to reason across different representations.

Reasoning With Measured Quantities

Quantitative reasoning, multiplicative reasoning, proportional reasoning, and reasoning with rational numbers are all related forms of reasoning with multiplicative structures (Sowder et al., 1998). Together, they form the bulk of mathematics content in the middle grades and reflect the interconnectedness within the content areas of middle grades mathematics (Sowder et al., 1998; Vergnaud, 1998). According to research in mathematics education, the process for developing additive reasoning and multiplicative reasoning is quite different. While additive reasoning can often be developed naturally and intuitively, the subsequent development of multiplicative reasoning often requires deliberate cultivation of deep conceptual understanding (Sowder et al., 1998). The transition from primarily focusing on additive reasoning in the primary grades to the emphasis of multiplicative reasoning thus poses particular challenges for middle grades learners. One such challenge is associated with reasoning with measured quantities, which is inherently different than numerical reasoning (Sowder et al., 1998). For example, quantitative reasoning often requires attending to units, and it is especially challenging for learners to operate with and coordinate intensive or composed units (e.g., five miles per hour) rather than extensive or singleton units (e.g., five miles, two hours).

According to Thompson (2011), a quantity is comprised of multiple components including an object, an attribute of an object, and a unit of measure that anticipates a number that is proportional to this unit of measure. Reasoning with measured quantities is different than reasoning with numbers alone. For example, $3 + 2 + 6 = 11$ requires only reasoning with numbers whereas reasoning with measured quantities might entail adding 3 dogs, 2 cats, and 6 lizards together to result in 11 animals. Quantitative operations are based on a situation and are not simply decontextualized numerical calculations, but numerical operations can be used to evaluate a quantity (Thompson, 1994). For example, a quantitative operation for the animal example will recognize that adding more animals will result in a greater number of animals, and it also involves coming to understand the process of creating a collective whole of animals from the constituent collections of animal types. Quantitative operations do not necessarily include numbers, and numerical operations do not necessarily include quantities.

There is much focus on numerical operations and algorithms in school mathematics, but the ability to follow a prescribed algorithm does not demonstrate quantitative reasoning. Students are reasoning quantitatively when they work with numbers in context and use appropriate operations across different situations. The quantities themselves and the relationships between them are created by an individual as they work through a problem (Moore, 2014), and what quantities and relationships a student knows should not be taken for granted (Izsák, 2003).

Not only students but also prospective and practicing teachers find quantitative reasoning challenging (Sowder et al., 1998). Teachers made similar errors and experienced similar struggles (e.g., having to coordinate two dimensions) as middle grades students (e.g., Simon & Blume, 1994; Sowder & Philipp, 1995). Therefore, Sowder et al. (1998) recommend that future teachers be provided with extensive opportunities to explore situations that require them to make sense of quantitative relationships and express them explicitly. Future teachers need such opportunities to help them form the habit of analyzing a situation quantitatively first before focusing on numbers and numerical relationships. This includes identifying meaningful quantities and estimating results with respect to their quantitative meaning.

In terms of developing mathematics language, Sowder et al. (1998) noted that future teachers have a strong tendency to reason exclusively in terms of numbers and numerical operations. In other words, they have mastered "the language of numbers and numerical operations" (Sowder et al., 1998, p. 135), but still need to develop the language of quantitative relationships because they need the latter to help their students make connections between arithmetic relationships and the situations they represent and evaluate. Sowder et al. (1998) also noted that while engaging in quantitative situations, future teachers might use everyday, nontechnical language to describe situations before attempting to use both conventional and unconventional notations to more formally express their reasoning about quantitative relationships.

Proportional Reasoning

Proportional reasoning is a mature mathematical way of thinking that consolidates many elementary concepts, generally emerges in middle grades, and continues to deepen and expand throughout later years (Lamon, 2012). Multiplicative and proportional reasoning are important because their development, or lack thereof, can greatly influence success for students in later mathematics (Beckmann, Izsák& Ölmez, 2015, Izsák, Kulow, Stevenson, Ölmez, & Beckmann, 2019, Lamon, 2007). This important area of mathematics continues to be investigated because as Touniaire and Pulos (1985) claimed, "it is both useful and difficult to master" (p. 181). Whereas ratios and proportions are a focus of middle-grades mathematics, they are challenging concepts for students and adults, including future and practicing teachers (Beckmann & Izsák, 2015; Behr et al., 1992; Izsák & Jacobson, 2017; Kaput & West, 1994; Karplus et al., 1983; Lamon, 2007; Tourniaire & Pulos, 1985).

Researchers associate reasoning proportionally with a thoughtful approach to contextual problems using common sense rather than blindly applying an algorithmic approach (Lamon, 2007). When reasoning with proportional problems, many students struggle because they have problems coordinating multiple quantities or use additive rather than multiplicative reasoning. Additionally, students may not recognize a fraction as a number, may have difficulty choosing which quantities to compare, and may struggle in determining the appropriate mathematical operation to use (Greer, 1992), thus they may focus on key words to guess at the operation. Students may not have developed fluency converting between fractions, decimals, and percentages and may not have developed an understanding that these values express a relationship of two numbers. Also, because linearity is the most intuitive relationship, students may assume this type of relationship exists when in fact it does not, and additionally they may mistake a linear relationship that does not go through the origin as proportional, both leading to incorrect conclusions. Whereas proportional reasoning clearly requires a deep, conceptual understanding of a myriad of mathematical objects such as fractions, quotients, ratios, and rates, too often, it is enacted procedurally by both students and teachers (Fisher, 1988; Harel & Behr, 1995; Izsák & Jacobson, 2017; Orrill & Brown, 2012) by applying the cross-multiplication algorithm.

Lobato and Ellis (2010) describe proportional reasoning as a complex process that requires essential understanding of the mathematics concepts involved, including equivalent ratios, ratios as composed units versus multiplicative comparisons, and the connections that link ratios, fractions, and quotients. Forging the essential understanding of proportionality calls for contextually meaningful and cognitively demanding tasks. For example, in a seminal study aimed at assessing proportional reasoning, Noelting (1980) used multiplicative comparison problems and had students consider the anticipated taste of a mixture of concentrated orange juice and water with varying amounts to determine which had a stronger orange taste.

Proportional reasoning problems traditionally are represented by graphs (e.g., a line), equations (e.g., $y = kx$, $a/b=c/d$), and tables (e.g., list of values for x and y). In addition, these problems are also presented with the use of a constant of proportionality or a scale factor, which can be seen geometrically when dilating an object. However, there are some other very useful representations available. To enhance proportional reasoning, many instructional materials for ratios and proportions include visual representations like double number lines, strip diagrams (tape diagrams), and 100 blocks to assist students in making sense of this content (Beckmann, 2018; Van de Walle et al., 2019).

METHODS

We analyzed transcripts from six future middle grades teachers who were interviewed in the fall of 2016 using open coding and the constant comparative method (Strauss & Corbin, 1998). These future teachers (FTs) were enrolled at a large public university in the southeast in the first of a two-semester sequence of mathematics content courses, the first with a focus on numbers and operations and the second with a focus on algebra. These semistructured interviews were video recorded, all written work created during the interview was scanned, and the interviews were transcribed verbatim. The data for this study is part of a broader study of future teachers' reasoning about multiplication and division, fractions, and proportional relationships.

This study considers two particular tasks, the Daffodil Problem and the Paint Problem. These problems were given to each of the six participants in the third interview. The interviews took place near the end of instruction focusing on multiplication and division, but before instruction on ratio and proportional relationships. Each interview was recorded with two video cameras in order to capture both a close-up view of the participant's work and a broader view including body language, mannerisms and facial expressions (Hall, 2000). Both authors watched the interviews, read the interview transcripts and cocreated lesson graphs (Seago et al., 2004), making analytic notes to document our interpretations of the FT's reasoning.

The Daffodil Problem

The Daffodil Problem was designed to target a sense of invariance (i.e., both quantities in a ratio change by the same factor) and investigate when using a math drawing can be helpful and when it might be a burden. The problem includes two parts, A and B given to the participants separately. The first part is simply a word problem about a gardening company:

> A gardening company plants 224 daffodils per acre. The company planted 392 daffodils at a park. How many acres of daffodils did the company plant?

The second part includes the same word problem and asks the participants to use the features of the two drawings below (see Figure 5.1) to solve this problem.

Figure 5.1

Visual Representation Provided for the Daffodil Problem, Part B

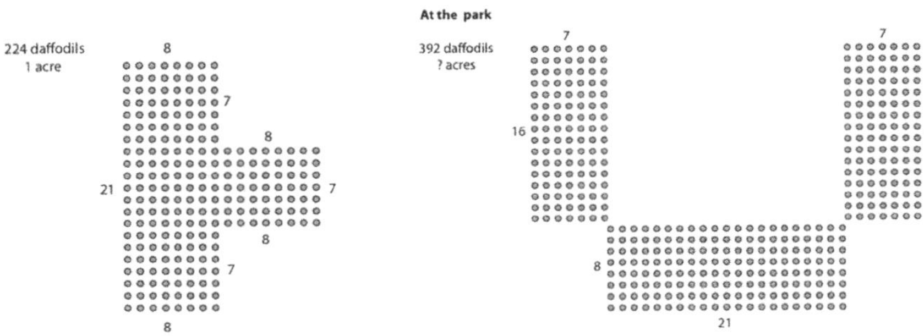

The Paint Problem

The Paint Problem was also designed to target a sense of invariance and investigate when using a math drawing can be helpful and when it might be a burden. Additionally, this problem aimed to uncover what aspects of the problem are important to show explicitly in the drawing. What makes the Paint Problem different than the Daffodil Problem is that the first part of the Paint Problem requested the participant to create a drawing rather than providing a drawing to the participant:

Consider the following information about a mixture of blue and yellow paint:

- The mixture consists of 3 parts blue and 5 parts yellow.
- Each part contains the same amount of paint.
- There are X gallons of blue paint and Y gallons of yellow paint in the mixture.

Please produce a math drawing that shows relationships among all the quantities. The second part asks the participants to write a mathematical expression that describes the relationships among the quantities shown in the drawing they produced.

RESULTS

The FTs worked with the Daffodil Problem for which they were provided a visual representation, and the Paint Problem for which they were asked to create a visual representation. Both problems specified an invariant ratio. In the Daffodil Problem, an additional measured quantity (an 8 x 7 plot) was introduced, whereas for the paint problem, two variables (X and Y) were introduced.

The Daffodil Problem

To solve the daffodil problem, it is necessary to use proportional reasoning. A number of daffodils per acre were given as a standard rate along with the number of daffodils in a particular park; however, the size of the park was unknown. As the FT's worked with the visual display provided in part B, more was revealed about their reasoning with visual representations, quantities, and proportions.

Division Versus Proportion: Part A of the Daffodil Problem

Overall, the FTs were able to solve Part A of the problem using either a division method or a proportion method. Most of the FTs (4 out of 6) chose division as opposed to a proportion since they had recently been exposed to instruction that focused on division and multiplication in the mathematics content course mentioned. Among the four FTs who chose to use division to solve this problem, all of them were able to identify 392 ÷ 224 as measurement division (how many groups) rather than partitive division (how many in each group). For example, Molly explained: "This is how many groups because we know how many are in each group. Because we have… so like a group is an acre, and we know that there are 224 daffodils in each acre." Both FTs who used a proportion (224/1 = 392/x) to solve the problem reasoned using either distance between daffodils or the space each daffodil occupied. For example, Susan explained,

> Well, I think about proportional reasoning. That's like how I see it because you have 224 per 1 acre, so then they're asking how many acres for 392 daffodils. So if … there are that many flowers and this amount of distance,

> there's going to be that amount of flowers in this amount of distance. Like there are going to be proportional amounts.

Similarly, Lily reasoned,

> Well, it just says that 224 daffodils per acre. And you know you have 392 daffodils, and you're trying to find out how many acres, so it's always that same number of daffodils per acre. So it's going to be like in the same space ... each daffodil. So it's going to take ... so like 1 would take the same amount. So that has to be the same.

Both FTs struggled to explain why the cross-multiplication algorithm works. Lily, for example, stated: "Because I know if you cross-multiply, you get an equation ... just what I have been taught."

When FTs were prompted to think whether they could use another method to solve the problem, five out of the six FTs were not able to identify another method. One FT (Susan) was able to reason that the problem could be meaningfully interpreted as either a division problem or a proportion problem.

Quantifying an Object

As we analyzed the transcripts, we use the term "plot" for each of the 7x8 rectangle of 56 daffodils (see Figure 5.2). These are the equal sized groups where there are four plots in an acre. Using a measurement unit of plots rather than daffodils allows participants to recognize that the daffodils planted at the park is equivalent to 7/4 acres without carrying out the numerical calculation of 392 ÷ 224. This is because there are 4 plots in an acre and 7 plots at the park.

Quantifying an object includes envisioning both an attribute to be measured and a unit of measure. Many of the FTs had difficulty in naming the quantity of a "plot" and this may have impacted their reasoning. Words such as rectangles, blocks, "smallers," boxes, thingy, and squares were used to identify the ¼ of an acre. Those FTs who identified a name and used it consistently provided more coherent and better coordinated explanations. Additionally, those who identified the number of daffodils (56) for this plot were able to create an equation.

Some of the participants struggled to keep up with the appropriate whole, and two FTs considered the plot to be 1/3 of the whole, as they were using the larger rectangular 8 x 21 block as the whole instead of this area plus a plot, which represents one acre. These FTs self-regulated and found the right relationship, one after receiving a scaffolding question from the interviewer. We consider this to be a constraint of reasoning with the visual display.

Figure 5.2

FTs Work Using Plots to Determine Acres

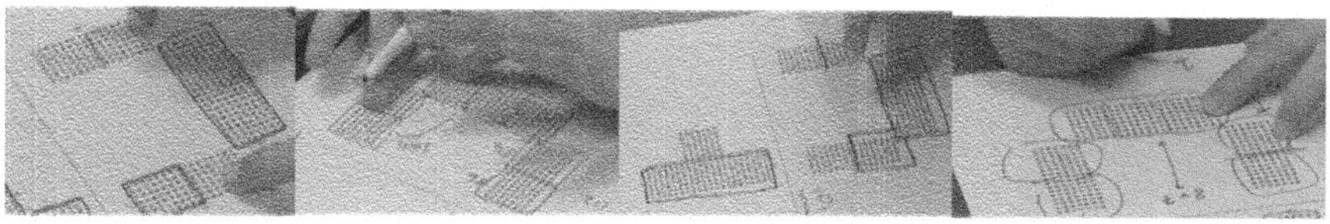

Finding Equivalence

Most of the participants began Part B of the Daffodil Problem by matching like size regions with more of a focus on shape or number. They generally found the whole acre in the park drawing, then determined the value of the "leftover" part, approximating the value as somewhere between one and two. All FTs gave their initial answer in the form of a mixed fraction (1 3/4) which has benefits of being able to see where it falls on the number line. However, it may be this proper form that is stressed in schools that limits their proportional reasoning and the ability to recognize the 7:4 ratio and apply it to see the 7/4 acres of daffodils that are planted at the park. However, with prompts from the interviewer, all of the FTs were able to convert the mixed fraction to the corresponding improper fraction.

When asked whether 329/224 is equivalent to 7/4 (i.e., invariance), all the FTs responded yes, but with varying degrees of confidence. Some struggled to produce the proportion (392/224 = 7/4). At times FTs used drawings to come up with relationships, but then did not acknowledge the usefulness of the drawing. For example, Lily was able to use the visual display to see the ratio of plots in an acre to plots in the park as 4:7 but claimed to "use my math rather than these drawings." Although Lily was unable to produce an equation for part B, which we contribute to her lack of understanding of the measurement attribute, she was able to speak to and represent the equivalence of the relationships (see Figure 5.3).

We also found that a more flexible view of the referent unit in this context, which was called the "base unit" in the mathematics content course, facilitated the identification of the equivalence relationship between the two ratios (i.e., 392:224 and 7:4). For instance, most (5 out of 6) of the FTs explained the subdivided acre with multiplication by retaining the base unit as a daffodil and redefining a group as a plot. However, Molly, who in our opinion had the most coherent explanation and worked flexibly throughout the problem, retained the acre as a group and redefined the base unit as a plot (she consistently called this a "square") as opposed to retaining the base unit as a daffodil (see Figure 5.4).

Figure 5.3

Lily's s Work Representing Equivalence of Fractions

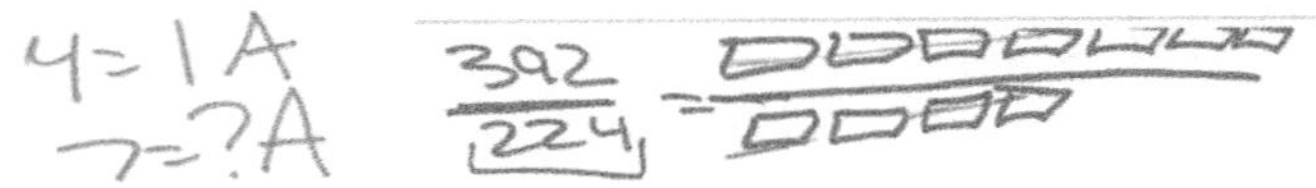

Figure 5.4

Molly's Work Identifying Plots ("squares") as the Base (referent) Unit

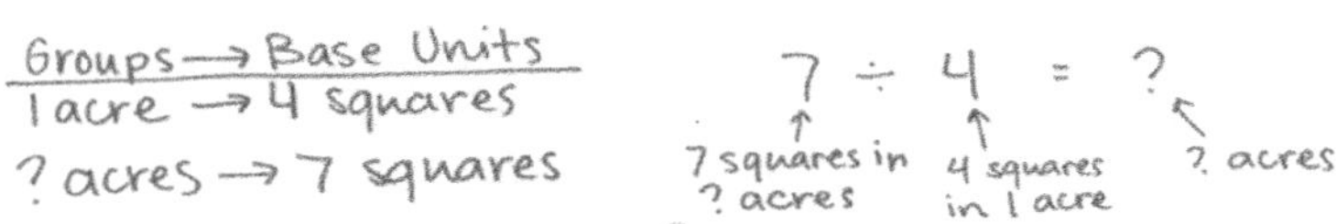

The Paint Problem

It is also necessary to use proportional reasoning when solving the paint problem. The paint problem includes a ratio of blue to yellow paint (3:5) and defined variables X and Y. The FTs were asked to create a drawing. The request for a math drawing is something that occurred frequently in their class and various understandings of the problem and the variables emerged.

Drawing, Labeling, and the Equal Sign

The FTs were not given a visual model for the Paint Problem. Instead, they were asked to produce their own models. Figure 5.5 shows a range of representations the FTs created. All the FTs produced an appropriate drawing for solving the Paint Problem, however two of the six FTs struggled to label their drawing appropriately. One of the FTs, Jack (see Figure 5.5B) used a strip diagram, which was recommended for similar problems in the mathematics content course previously mentioned. Although the problem stated there are X gallons of blue paint and Y gallons of yellow paint in the mixture, he labeled the blue paint as 3X and the yellow paint as 5Y. Jack indicated that he initially thought about this as a ratio of 3X:5Y and that the equation that he wrote, 3X= 5Y, meant the same thing. When

Figure 5.5

The Future Teachers' Visual Representations of the Paint Problem

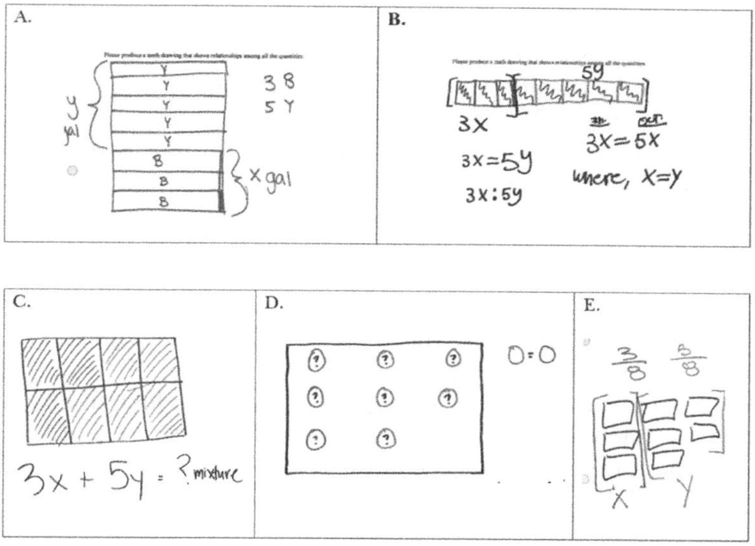

asked where he saw X and Y in his drawing, Jack recognized each part was an equal size but simultaneously referred to the parts as both X and Y:

> The X I drew from the first bullet point where it says we had 3 blue parts in our mixture. So I … and then it says we had X gallons of blue paint. So I just made a choice that 1 part blue could be equal like to 1 gallon. So I said that for very blue part would be how many … half of the 3X and then same logic for the 5Y. We have 5 yellow parts. So I just assumed that if all sizes were equal that I can have 5Y. But since all sizes are equal I could also just say … and X equals 5X where … X equals Y since the size does not matter since they are the same.

Jack struggled to identify and label what constituted one part. He stated there were X gallons of blue paint; however, he did not clearly relate the variable X to the entire three parts of the blue paint. He also used the equal sign in a variety of ways. In stating that 3X : 5Y and 3X = 5Y had equivalent meanings, Jack was not using the equal sign in a conventional way where the right-hand side is equivalent to the left-hand side. Another unconventional use of the equal sign is 3X = 5X, which Jack explains as "it's more a function where this is our input and this would be the output."

Only his last equation X = Y includes a conventional use of the equal sign which he identifies as meaning they are the same size.

Similarly, Nina connected the variables X and Y with a single blue part and yellow part respectively rather than with multiple parts. Initially, as shown in Figure 5.5C, she drew eight equal size parts, colored 3 blue and 5 yellow, and before writing any equation she questioned how to label her diagram: "I guess I could just write an X on them, but I don't know how to get an X on all 3 of them or a Y on all 5 of these." When asked how she was thinking about the X and the Y, Nina replied, "some form of X plus Y is going to equal the paint mixture" and hesitantly suggested 3X + 5Y (see Figure 5.5C). When asked where she saw 1X and 1Y in her drawing, Nina identified each as the amount of paint in one part, which is consistent with the equation she wrote. Thus, the letters X and Y were not seen as variables that represented the quantities (in gallons) of the blue paint and the yellow paint respectively. Instead, they were treated as labels or units, that is, substitutions for the words "blue paint" and "yellow paint."

Unlike Jack, Nina seemed to have the correct understanding of the quantitative relationships in the problem and used the equal sign in a conventional way that signified equivalence. She explained that if 3X + 5Y represented the total mixture, then X and Y represented the number of gallons in one part. Nina's difficulty stemmed from her struggle to connect the quantities in her equation with her drawing. While Jack struggled to work with measured quantities and used an inconsistent interpretation of the equal sign to communicate, Nina's work reflected her lack of comfort with using drawings as a tool to aid mathematical understanding.

When both Nina and Jack were asked to consider X to be the total amount of blue paint and Y to be the total amount of yellow paint, they each produced normative representations (see Figure 5.6).

Creating Equations

All six FTs were eventually able to label their drawings appropriately, however only four of the six FTs created at least one of the following equations that represent the situation appropriately:

$$Y = 5/3X \text{ or } X = 3/5Y \qquad (1, 2)$$

$$Y/X = 5/3 \text{ or } X/Y = 3/5 \qquad (3, 4)$$

$$X/3 = Y/5 \qquad (5)$$

Two FTs either did not generate an equation (Molly) or generated an incorrect equation (Lily). The equations produced by the FTs, correct

Figure 5.6

Nina's and Jack's Second Representation of the Paint Problem

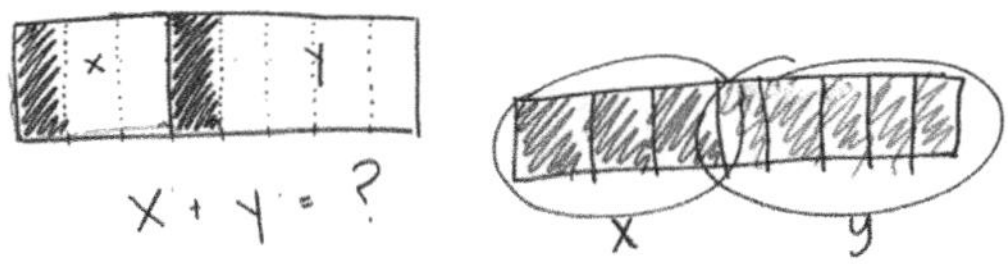

or incorrect, revealed rich underlying conceptual thinking, which are analyzed below.

Multiplicative Comparison: The Case of Jack and Lily. Despite his initial struggle, Jack was able to produce a correct drawing and labeling. He also provided a pair of correct equations Y = 5/3X and X = 3/5Y (Equations 1, 2). When asked to explain, he stated:

> I'm kind of thinking of these two as kind of the same as … for however much amount of blue paint we have we'll have … if we have 3 of blue paint, we're going to 5Y or factors thereof. So if we reduce that just to X, we have 5/3 the amount of yellow paint to blue paint.

Jack was able to use multiplicative comparison to correctly relate the amount of blue paint (X) to the amount of yellow paint (Y), however his verbal explanation still included the representation 5Y rather than Y. This was consistent with his initial struggle in labeling his drawing and similar to Nina's initial work which treated Y as a label or unit that was attached to 5 instead of a variable that represented the quantity of the yellow paint.

Lily also used the multiplicative comparison strategy, but produced an equation of X = 1 2/3Y, rather than Y = 1 2/3X. She made the classic "reversal error," which was documented by researchers such as Clement and colleagues (e.g., Clement, 1982; Clement et al., 1981). When asked to justify the meaning of the equation, Lily explained: "but it's like for every 1X you have 1 and 2/3Y ... I mean it would … it's just be like X equals 1 and 2/3Y." Unlike Nina who initially treated X and Y as labels or units instead of variables, Lily seemed to have identified the variables and a proportional relationship between the two types of paint. But she was not able use operative reasoning to translate her understanding of the variables to an appropriate relationship. Operative reasoning allows recognition that since there is more yellow paint than blue paint, the amount of blue paint, X must be multiplied by a number greater than 1 (1 and 2/3 in this case) to produce the amount of yellow paint, Y. In essence, this is answering the question, "Which quantity is larger, X or Y?" and interrogating the final answer by asking, "If Y is larger and is multiplied by a number that is greater than 1, can this expression produce the smaller quantity X?"

Set up a Proportion: The Case of Susan. Susan was the only one who set up a proportion or an equality between two ratios, Y/X = 5/3 (Equations 3, 4). When pressed to explain how to make sense of her equation, that is, why on the left side, Y and X represented gallons, but on the right side, 5 and 3 were parts, she offered the following explanation:

> Well, I'm comparing ... these are parts, but they don't have like a ... like a unit, I guess, assigned to them. So the X and Y are in terms of gallons. So ... but because each of the parts are the same size, we can compare like 5 yellow parts to 3 blue parts because they're each the same size. And then that allows us to ... talk about it in terms of gallons by talking about Y gallons and X gallons.

Susan's reasoning resembled a composed unit strategy in that she joined 3 parts of blue paint and 5 parts of yellow paint to form a composed unit or a "batch." When asked whether she would change her equation if each part of the paint is stretched or compressed, Susan demonstrated a sense of invariance as she insisted that the relationship Y/X = 5/3 should remain regardless of the size or quantity of one part of paint.

Ratios as Fractions and Quotients: The Case of Nina and Michelle. Nina and Michelle were the only two FTs who were able to come up with two equations (1, 2) and (5). Both use the multiplicative comparison strategy, which was the most frequently used strategy among the six FTs (four of the six used this strategy). What was unique about Nina and Michelle was that both produced the equation X/3 = Y/5 and interpreted the ratios X/3 and Y/5 as fractions and then also as division. For example, Nina explained the equivalence relationship in the proportion as "equal fractions":

> I know that X ... like 1/3 of X ... is the same size of 1/5 of Y. So that helps me know like if I have this much paint I know how much yellow paint I am going to have.... It's like 1/3 of X, so X partitioned into 3 pieces. It would be 1 of those pieces.... So Y partition into 5 pieces, and it's 1 of those pieces [pointing to her drawing].

When asked whether she saw X/3 and Y/5 as divisions, Nina's initial response was "no, I'm thinking of equal fractions." But eventually, Nina was able to see the two ratios in the equation as two divisions (or quotients) as well. She was also able to correctly identify the division as a "how many in 1 part" or partitive division rather than measurement division. This demonstrates a sense of proportional reasoning that recognizes each ratio as a quotient and incorporates division where each quantity is considered and compared as an equal size part.

Variables and Inequalities: The Case of Molly. Molly used a question mark instead of a conventional letter variable to represent the amount of gallons in one part of paint. She produced two equations X = 3(?) and Y = 5(?) relating the equal size part of any color paint (?) to the total blue paint (X) and yellow paint (Y). However rather that generating an equation to relate the blue and yellow paint, she produced a correct inequality, X < Y. If instead P were assigned as a variable to the equal size part, Molly might be able to rewrite her equations as X = 3P and Y = 5P, then either an equation of X/3 = Y/5, Y = 5/3(X), or X = 3/5(Y) could be created with fairly simple algebraic manipulations. It is difficult to tell whether the use of the "?" rather than a traditional letter variable prevented Molly from recognizing the direct relationship between X and Y and forming an equation.

DISCUSSION

In this study, six future mathematics teachers explored proportional reasoning using visual representations. Visual representations afforded a wider range of thinking than word problems alone. For example, Part A of the Daffodil Problem was presented in words with no visual representation and although all the future teachers in our study were able to produce the correct answer, the word problem alone did not afford diverse ways to think about the problem. The future teachers either chose to use a division method or a proportion method to solve the problem, and all except for one were not able to think of another method. The future teachers who used a proportion method applied the cross-multiplication algorithm but were not able to explain why it was a suitable strategy. In Part B of the Daffodil Problem, the future teachers were presented with a visual model, which grouped the daffodils into plots and broadened the ways the future teachers thought about the problem and forced them to reason about the quantities and the equivalence relationship in the proportion 392/224 = 7/4. For example, most FTs were able to see or interpret 392/224 as either a ratio, a fraction, or a quotient. They were also challenged to reconsider the referent unit they chose for the first part of the problem. Our results were consistent with those of Lamon (1993) who argued that word problems alone do not necessarily elicit reasoning, the problem type and how the problem is phrased affect the level of reasoning. Our results also show that mathematics can be seen as a language and a science of patterns. By providing a visual model, we oriented future teachers to the contextual and visual cues (Wakefield, 2000).

In addition, the visual representations helped reveal future teachers' thinking, and made their thinking visible. For example, the Paint Problem confirmed that multiplicative and proportional reasoning tend to be challenging for both students and adults, including future teachers (Sowder et al., 1998). Creating an appropriate representation for a seemingly straightforward word problem is not a trivial task. We observed that two of the six future teachers struggled with creating such a representation, and their incorrect representation led to incorrect mathematical expressions or equations. Our findings indicate that simply asking teachers and students to provide a drawing is often not enough in promoting conceptual understanding or reasoning, it is also important to provide scaffolding through the process, especially for properly labeling the variables in a drawing.

We suspect that the six future teachers as a group had more difficulty with the Paint Problem for two reasons. First, the Paint Problem required them to create a representation or visual model of their own, whereas for the Daffodil Problem, we provided a drawing to the future teachers. And earlier, we discussed that creating a representation for a word problem is not a trivial task and this approach of making thinking visible is often not required in school mathematics. Second, although both problems required future teachers to reason beyond numerical operations to engage with measured quantities and thus reason quantitatively, the Paint Problem, unlike the Daffodil Problem, introduced variables X and Y. In the Daffodil Problem, only numbers were introduced, and no variables were specified. The Paint Problem went beyond the traditional missing-value and multiplicative comparison types of proportional reasoning problems (Sowder et al., 1998) and introduced variables (or algebraic symbols) which clearly posed more challenges to the future teachers.

Our results were consistent with those of the Professor Problem (e.g., Clement, 1982; Clement et al., 1981). Clement and colleagues asked undergraduate students who were enrolled in calculus to solve the following a simple word problem:

> There are six times as many students as professors at this university. Use S for the number of students and P for the number of professors.

To their surprise, they found 37% of science-oriented students and 57% of nonscience majors in their sample missed this simple problem. Among the errors, a reversal error $6S = P$ (instead of the correct response $6P = S$) was the most common. After a detailed analysis of their clinical interview protocols, Clement and colleagues concluded that this reversal error was not simply a careless mistake, but "a self-generated, stable, and persistent misconception concerning the meaning of variables and equations" (Clement et al., 1981, p. 289). In addition, Rosnick and Clement (1980) found that this error did not occur only when a problem was presented in words, it was quite robust and persisted in translations from pictures to equations as well. Clement and colleagues further concluded that the reversal error had two main sources: syntactic translation and static-comparison. Syntactic translation is a “literal, direct mapping of the words of English into the symbols of algebra” or “word order matching” (Clement et al., 1981, p. 288). No FTs in our sample used this syntactic translation scheme, as they all were able to create an appropriate drawing to represent the problem that correctly identified the amount of the yellow paint was greater than that of the blue paint. Static comparison on the other hand is a mistake that is still made even when the problem solver correctly identifies which of the two quantities is bigger. Clement and colleagues attributed this mistake to misunderstanding of symbols such as X and Y as labels or units instead of variables that represent quantities. For example, in our sample, Nina and Jack both made similar mistakes by attaching X (gallons of blue paint) to 3 and Y (gallons of yellow paint) to 5, According to Clement et al. (1981), students who used static comparison scheme to set up an incorrect equation, in our case, $3X = 5Y$, often misinterpret the equal sign as a comparison or association, instead of a precise equivalence. This coincides with Jack’s multiple uses of the equal sign.

To correctly express the Paint Problem or the Professor Problem, one must be able to engage in operative translation, that is, to realize which quantity is larger (or smaller) and what operation needs to be performed on the large number (smaller number) to produce the smaller number (large number). For example, in our case, the amount of the yellow paint was the larger of the two quantities. Therefore, in order to perform an operation on the smaller quantity, the amount of blue paint (represented by the variable X), to produce a number that is the same as the amount of the yellow paint (represented by the variable Y), X must be multiplied by 1 2/3, a multiplier greater than 1. In our sample, Lily was not able to correctly perform this operative translation and therefore, produced the reversal error even though she identified the correct proportional relationship between X and Y in her drawing. All in all, our study replicated the results obtained from the research done by Clement and colleagues four decades ago and showed one more time that translating across different representations (e.g., words, pictures, and algebraic notations) can be extremely challenging even for mathematically focused students such as FTs. Clement et al.’s (1981) conclusion that “translation skills, a long-neglected component of mathematical literacy” (p. 289) still seems to ring true today.

Finally, although the visual representations in this study facilitated and broadened future teachers’ proportional reasoning, they did not necessarily lead to the automatic use of formal mathematical language or terminology. For example, in solving the Daffodil Problem, no one in the study used the word “equivalence” or “invariance” to describe $392/224 = 7/4$. Instead, some future teachers used words such as “same” or “similar” to describe the two ratios. Our results were consistent with Sowder et al.'s (1998) observation that future teachers might choose to use everyday, nontechnical language to reason through a problem instead of relying on formal mathematical language. Another example is: When quantifying the object, "plot," in the Daffodil Problem, future teachers' choice of language was quite idiosyncratic and sometimes had a lack of precision, for example, calling a 7*8 rectangular plot a "square."

CONCLUSION

Through working with future middle grades teachers, we recognize their desire to understand and figure things out in an efficient manner. They are in the process of developing their vocabulary, and mathematical terms not used in everyday language, or used in a different manner, are challenging. Their explanations may sometimes seem incorrect or incomplete because of imprecise vocabulary or connections, but they seem to have logical underpinnings based on connections they have made or are forming. Providing learners with or requesting them to create representations helps to make their thinking visible and allows them to formulate more precise explanations of the quantitative relationships. The wording of questions in mathematics can be challenging and understanding what the questions are asking can be difficult. We should allow students to create their own representations and guide them in improving the precision of these representations. Allowing students to reason across multiple representations allows them to better understand problems, create appropriate connections, and reason quantitatively with the constituent components.

REFERENCES

Arcavi, A. (2003). The role of visual representations in the learning of mathematics. *Educational Studies in Mathematics, 52*(3), 215–231. https://doi.org/10.1023/A:1024312321077

Beckmann, S. (2018). *Mathematics for elementary teachers, with activities* (5th ed.). Pearson.

Beckmann, S., & Izsák, A. (2015). Two perspectives on proportional relationships: Extending complementary origins of multiplication in terms of quantities. *Journal for Research in Mathematics Education, 46*(1), 17–38. https://doi.org/10.5951/jresematheduc.46.1.0017

Beckmann, S., Izsák, A., & Ölmez, İ. B. (2015). From multiplication to proportional relationships. In X. Sun, B. Kaur, J. Novotna (Eds.), *Conference proceedings of ICMI Study 23: Primary mathematics study on whole numbers* (pp. 518–525). University of Macau.

Behr, M. J., Harel, G., Post, T. R., & Lesh, R. (1992). *Rational number, ratio, and proportion.* In D. A. Grouws (Ed), *Handbook of research on mathematics teaching and learning* (pp. 276–295). Macmillan.

Boaler, J., Chen, L., Williams, C., & Cordero, M. (2016). Seeing as understanding: The importance of visual mathematics for our brain and learning. *Journal of Applied & Computational Mathematics, 5*(5), 1–6. https://doi.org/10.4172/2168-9679.1000325

Clement, J. (1982). Algebra word problem solutions: Thought processes underlying a common misconception. *Journal for Research in Mathematics Education, 13*(1), 16–30.

Clement, J., & Lochhead, J., & Monk, G. S. (1981). Translation difficulties in learning mathematics. *The American Mathematical Monthly, 88*(4), 286–290.

Dreher, A., & Kuntze, S. (2015). Teachers' professional knowledge and noticing: The case of multiple representations in the mathematics classroom. *Educational Studies in Mathematics, 88*, 89–114. https://doi.org/10.1007/s10649-014-9577-8

Fisher, L. C. (1988). Strategies used by secondary mathematics teachers to solve proportion problems. *Journal for Research in Mathematics Education*, 157–168.

Greer, B. (1992). Multiplication and division as models of situations. In D. A. Grouws (Ed.), *Handbook of research on mathematics teaching and learning* (pp. 276–295). Macmillan.

Hall, R. (2000). Videorecording as theory. In R. Lesh (Ed.), *Handbook research data design in mathematics and science education* (pp. 647–664). Lawrence Erblaum Associates.

Harel, G., & Behr, M. (1995). Teachers' solutions for multiplicative problems. *Hiroshima Journal of Mathematics Education, 3*, 31–51.

Izsák, A. (2003). "We want a statement that is always true": Criteria for good algebraic representations and the development of modeling knowledge. *Journal for Research in Mathematics Education, 34*(3), 191–227. https://doi.org/10.2307/30034778

Izsák, A., Kulow, T., Stevenson, D., Ölmez, İ. B., & Beckmann, S. (2019). Using coordinated measurement with future teachers to connect multiplication, division, and proportional relationships. *Mathematics Teacher Educator, 8*(1), 49–75.

Izsák, A., & Jacobson, E. (2017). Preservice teachers' reasoning about relationships that are and are not proportional: A Knowledge-in-Pieces Account. *Journal for Research in Mathematics Education, 48*(3), 300–339. https://doi.org/10.5951/jresematheduc.48.3.0300

Kaput, J. J. (1992). *Technology and mathematics education.* In D. A. Grouws (Ed.), *Handbook of research on mathematics teaching and learning* (pp. 515–556). Macmillan.

Kaput, J. J., & West, M. M. (1994). Missing value proportional reasoning problems: Factors affecting informal reasoning patterns. In G. Harel & J. Confrey (Eds.), *The development of multiplicative reasoning in the learning of mathematics* (pp. 235–287). State University of New York Press.

Karplus, R., Pulos, S., & Stage, E. K. (1983). Early adolescents' proportional reasoning on 'rate'problems. *Educational Studies in Mathematics, 14*(3), 219–233. https://doi.org/10.1007/BF00410539

Lamon, S. (1993). Ratio and proportion: Connecting content and children's thinking. *Journal for Research in Mathematics Education, 24*(1), 41–61.

Lamon, S. J. (2007). Rational numbers and proportional reasoning: Toward a theoretical framework for research. In F. K. Lester (Ed.), *Second handbook of research on mathematics teaching and learning* (Vol. 1, pp. 629–667). Information Age Publishing.

Lamon, S. J. (2012). *Teaching fractions and ratios for understanding: Essential content knowledge and instructional strategies for teachers* (3rd ed.). Routledge. https://doi.org/10.4324/9780203803165

Lobato, J., & Ellis, A. B. (2010). Developing essential understanding of ratios, proportions, and proportional reasoning for teaching mathematics in grades 6-8 (R. J. Charles, Ed.). In R. M. Zbiek (Series Ed.), *Essential understanding series*. National Council of Teachers of Mathematics.

Moore, K. C. (2014). Quantitative reasoning and the sine function: The case of Zac. *Journal for Research in Mathematics Education, 45*(1), 102–138. https://doi.org/10.5951/jresematheduc.45.1.0102

National Council of Teachers of Mathematics. (2014). *Principles to actions: Ensuring mathematical success for all.*

National Governors Association Center for Best Practices & Council of Chief State School Officers. (2010). *Common Core State Standards for Mathematics.* http://www.corestandards.org/assets/CCSSI_Math%20Standards.pdf 20

Noelting, G. (1980). The development of proportional reasoning and the ratio concept part II: Problem-structure at successive stages; problem-solving strategies and the mechanism of adaptive restructuring. *Educational Studies in Mathematics, 11*(3), 331–363.

Orrill, C. H., & Brown, R. E. (2012). Making sense of double number lines in professional development: Exploring teachers' understandings of proportional relationships. *Journal of Mathematics Teacher Education, 15*(5), 381–403.

Riccomini, P. J., Smith, G. W., Hughes, E. M., & Fries, K. M. (2015). The language of mathematics: The importance of teaching and learning mathematical vocabulary. *Reading & Writing Quarterly, 31*(3), 235–252. https://doi.org/10.1080/10573569.2015.1030995

Rosnick, P., & Clement, J. (1980). Learning without understanding: The effect of tutoring strategies on algebra misconceptions. *The Journal of Mathematical Behavior, 3*(1), 3–27.

Seago, N., Mumme, J., & Branca, N. (2004). *Learning and teaching linear functions: Video cases for mathematics teacher professional development. 6–10*. Heinemann.

Shulman, L. (1986). Those who understand: Knowledge growth in teaching. *Educational Researcher, 15*, 4–14.

Simon, M. A., & Blume, G. W. (1994). The troublesome idea of "ratio": Mathematical modeling as a component of understanding ratio as measure: A study of prospective elementary teachers. *Journal of Mathematical Behavior, 13*, 183–197.

Sowder, J., Armstrong, B., Lamon, S., Simon, M., Sowder, L., & Thompson, A. (1998). Educating teachers to teach multiplicative structures in the middle grades. *Journal of Mathematics Teacher Education, 1*, 127–155.

Sowder, J. T., & Philipp, R. A. (1995). The value of interaction in promoting teaching growth. In J. T. Sowder & B. P., Schappelle (Eds.), *Providing a foundation for teaching mathematics in the middle grades* (pp. 223–250). SUNY Press.

Strauss, A., & Corbin, J. (1998). *Basics of qualitative research: Techniques and procedures for developing grounded theory*. SAGE.

Thompson, P. W. (1994). The development of the concept of speed and its relationship to concepts of rate. In G. Harel & J. Confrey (Eds.), *The development of multiplicative reasoning in the learning of mathematics* (pp. 179–234). SUNY Press.

Thompson, P. W. (2011). Quantitative reasoning and mathematical modeling. In L. L. Hatfield, S. Chamberlain & S. Belbase (Eds.), *New perspectives and directions for collaborative research in mathematics education. WISDOMe Mongraphs* (Vol. 1, pp. 33–57). University of Wyoming.

Tourniaire, F., & Pulos, S. (1985). Proportional reasoning: A review of the literature. *Educational Studies in Mathematics, 16*(2), 181–204. https://doi.org/10.1007/PL00020739

Van de Walle, J. A., Karp, K., Bay-Williams, J. M., & Wray, J. A. (2019). *Elementary and middle school mathematics: Teaching developmentally* (10th ed.). Pearson.

Vergnaud, G. (1988). Multiplicative structures. In J. Hiebert & M. Behr (Eds.), *Number concepts and operations in the middle grades* (pp. 141–161). Erlbaum and Reston, VA: National Council of Teachers of Mathematics.

Wakefield, D. V. (2000). Math as a second language. *The Educational Forum, 64*(3), 272–279. https://doi.org/10.1080/00131720008984764

CHAPTER 6

PATTERNS AND RELATIONSHIPS WITHIN RATIO CONTEXTS

Students' Emerging Ideas Through Ratio Tables

Sinem Sözen Özdoğan, Didem Akyüz, and Michelle Stephan

INTRODUCTION

Mathematics as "a science of patterns" is a famous metaphor to describe mathematics (see Mulligan & Mitchelmore, 2009; Resnik, 1982, 2005; Van de Walle et al., 2016; Wittman, 2005). Furthermore, understanding the pattern as a mathematical entity and a tool within different mathematical contexts plays a vital role in teaching and learning mathematics. Patterns are addressed in the context of algebraic thinking explicitly; however, it is possible to identify them in other contexts as well.

The Common Core State Standards emphasize the importance of finding patterns in one of the eight mathematical practices: "look closely to discern a pattern or structure" (CCSSO, 2010). Working with them in different mathematical contexts, students explore generalizations, relationships, transformations, and especially the existence of rules (Billstein et al., 2013; Burns, 2015). Therefore, realizing the relationship between

Mathematics as the Science of Patterns:
Making the Invisible Visible to Students Through Teaching, pp. 99–125

the structures of mathematics and modeling them may provide a way for the learning of mathematics.

Connecting mathematics with other mathematical concepts, different representations, real-life situations, and various disciplines, is an essential skill to be developed not only by the students but also by the teachers (Kaput, 1994). Pattern recognition as structures consists of *identifying units of repetition* and *constructing composite units* (Steffe, 1994), which can also be considered as the fundamentals of multiplicative reasoning in ratios (Battista & van Auken Borrow, 1995; Papic et al., 2011; Stephan et al., 2015).

Encouraging using patterns within different mathematical concepts was suggested by incorporating ideas such as *unit of repeat* (Papic et al., 2011; Wittman, 2005) and *growing patterns* (Warren & Cooper, 2008). Several studies have explored instruction embodying connections between ratio and pattern concepts within the school curriculum in terms of multiplicative structures (see Mulligan & Mitchelmore, 2009; Wittman, 2005). In this respect, making connections between patterns and multiplicative reasoning has formed the basis for designing new learning environments that are compatible with research-based instructional principles for mathematical content.

Based on the related literature, the current study describes a learning environment in which students shape their understanding of ratios by interacting with their peers to create meaningful strategies as they develop and reason with ratio tables (Dole, 2008). Table utility to generate functional thinking has been mentioned in several sources (Markworth, 2012; Van de Walle et al., 2016). Thus, a ratio table can be a base for the development of emerging ideas for both ratio and patterns. In our investigation, the process through which students internalized multiplicative reasoning within a ratio table grew out of students' patterning strategies. Therefore, the research question is:

> What are the emerging ideas of the seventh-grade students that are characterized as number pattern utility while developing ratio tables in the context of teaching ratio and proportions within a classroom teaching experiment?

THEORETICAL FRAMEWORK

In order to explain the genesis of the research question, we elaborate on the theoretical background that underlay our framework. Within this framework, common points related to patterns and ratio reasoning are briefly presented. Second, the ratio table as a reasoning tool is discussed with its capabilities and limits. The instructional sequence is described thirdly to provide the teaching-learning environment in detail.

Connections Between Reasoning With Patterns and Ratios

Reasoning with patterns is considered significant for developing various mathematical skills and learning areas such as algebraic and functional thinking (Papic et al., 2011; Warren & Cooper, 2007, 2008). Number patterns and patterns in operations help the development and improvement of the fundamental mathematical skills of the children from kindergarten to secondary schools (Van de Walle et al., 2016).

Moreover, students who participated in pattern activities in the early grades perform better in mathematics in the following years (Papic, 2007). Therefore, reasoning with patterns is suggested to be infused in the school curriculum (Papic, 2007; Papic et al., 2011; Warren & Cooper, 2007, 2008).

Proportional thinking is another significant mathematical concept that takes place in the middle-grade mathematics curriculum as proposed by Inhelder and Piaget (1958) in the formal operational stage of the children's cognitive development. Furthermore, it affects students' higher education levels as well as an understanding of other disciplines (Howe et al., 2011; Lachance & Confrey, 2002; Lamon, 2012). It involves comparing two rational quantities through multiplicative relationships and includes rate, ratio, and proportion concepts in the curriculum. Comprehensive studies regarding rate, ratio, and proportional reasoning reveal that there is still a need to investigate further proportional reasoning and its acquisition (Behr et al., 1992; Lamon, 2012; Tourniare & Pulos, 1985).

Proportional reasoning involves comparing two rational quantities through multiplicative relationships and includes rate, ratio, and proportion concepts in the curriculum. Comprehensive studies regarding rate, ratio, and proportional reasoning reveal that there is still a need to investigate further proportional reasoning and its acquisition (Behr et al., 1992; Lamon, 2012; Tourniare & Pulos, 1985).

Patterns and ratio concepts involve multiplicative structures in essence. Studies related to ratio concepts emphasize making connections with decimals, fractions, and rational numbers (Ben-Chaim et al., 1998; Howe et al., 2011; Lachance & Confrey, 2002). On the other hand, there are relevant studies for patterns that connect ratio and pattern reasoning, although they mostly focus on young children's knowledge and skills to develop patterns (Papic & Mulligan, 2007).

Warren and Cooper have published various studies investigating pattern understanding in young children. In one of their studies, they examined ways of supporting students' understanding of ratio with the help of 9-year-old students' knowledge of repeating patterns (Warren & Cooper, 2007). They studied the activities, such as the transformation of repeating patterns to ratios and recording the data on a table. They also emphasized repeating patterns to introduce the ratio concept, utilizing tables to organize the data, and representing units.

Similarly, Mulligan and Mitchelmore (2009) described the relationship between patterns and ratios for the early childhood mathematics terms of "repetition, grouping, partitioning and unitis[z]ing, and all require spatial structuring in visualis[z]ing and organis[z]ing their structure" (p. 38). These processes are crucial for the understanding of ratio and proportion in middle grades (Lamon, 2012).

Furthermore, ratio tables may represent the tables suggested by pattern-related articles (Markworth, 2012; Van de Walle et al., 2016). In these tables, there are stages (Markworth, 2012) or steps (Van de Walle et al, 2016) that describe the structure of the developing geometrically growing pattern. After students had developed an understanding of the pattern, the transformation of the data to the table was suggested to support generalization and explicit functional relationships (Markworth, 2012).

Stage or step is written as one of the labels in the first row or first column. The other label generally includes the total number of the figures or the number of a different numerical variable. In this kind of table, the focus is on more than a multiplicative relationship between numbers and the formula for the generalization of that pattern. It was suggested that students investigate the link between the pictorial representation of the units and the structure of the number system, involving at least two units (Warren et al., 2012), and students found it challenging to find multiplicative structures (Lamon, 2012; Singh, 2000). The ratio table has been a tool for exploring multiplicative structures, but students tend to explore this relationship by themselves.

Ratio Table as a Tool to Teach Proportional Reasoning

A ratio table can be considered as an organized digital or manual spreadsheet tool in which there are at least two rows and two columns to display the quantities in a problem situation in a meaningful order. There are several characteristics of the ratio table described by researchers (Broekman et al., 2000; Dole, 2008; Middleton & van den Heuvel-Panhuizen, 1995) below:

1. The table consists of cells in which the numerical values are filled according to a rule between the numbers in a ratio problem.
2. The labels are written in the first row or first column to identify what the numbers in the cells describe.

3. Horizontal or vertical relationships between the cells in the ratio table reflect the same numerical consistency.
4. It is a tool to encourage students to use any multiplication or division strategy (build-up strategies as well) provided that the ratio between the cells is preserved.

These rules either can be determined by the students or can be explored by the students during instruction (Sozen Ozdogan et al., 2019). Furthermore, as Dole (2008) highlighted, students are free to design their ratio tables. They may create various ratio tables for the same activity as long as they obey the rule reflecting an order of equivalent fractions between the vertical or horizontal cells. Students can explore these characteristics of the ratio table based on their experiences with the tasks and based on the discussion with their peers.

Ratio and Proportion Instructional Sequence

The ratio and proportion instructional sequence used in this study was developed by Stephan and colleagues (2015). It was designed based on the realistic mathematics education (RME) theory, which offers a realistic environment that encourages students to develop mathematical structures. The concept of *realistic* has been defined as situations that students can imagine rather than only those that students have experienced (van den Heuvel-Panhuizen & Drijvers, 2020).

The sequence was designed according to eight learning outcomes (i.e., *big ideas)* and included the activities, mathematical tools, and possible topics of discourse (Stephan et al. 2015). Student activities involve large numbers and decimals, and an unfamiliar order of missing value and comparison problems (Tourniare & Pulos 1985). The ratio table was the suggested mathematical tool to guide students' organizations of the data based on the challenge in the given task. Each task involves missing value problems or comparison problems along with a rule of the ratio.

For example, for the Rule 1, a teacher can take care of 3 infants at a daycare; students may be asked to determine how many teachers are needed to care for 9, 12, and 33 infants. Problems such as these encouraged students to develop strategies for finding equivalent ratios while maintaining an invariant relationship between the two quantities in the ratio; therefore, they promoted proportional reasoning (Abrahamson & Cigan, 2003; Ercole et al., 2011; Middleton & van den Heuvel-Panhuizen, 1995).

A typical class period was organized into a launch-explore-discussion (LED) cycle within the instructional sequence (Stephan et al,, 2016). The teacher and participant-observer (the researcher) started the day with the *Launch* session, reminded students of the activity they covered in the previous lesson, and posed a new problem or set of problems for exploration. In the *Explore* session, students worked on the problems individually or with their partners and tried to solve problems in their way.

The role of the teacher was to allow students to engage in mathematical situations that revealed their way of mathematical thinking. After exploration, students were encouraged to share their solution strategies and to discuss their peers' questions with the class. In whole-class discussion sessions, students compared and contrasted solution strategies to find the "best" solutions. Throughout this process, the discourse generated from whole-class discussion gave prominence to students' ways of thinking.

Furthermore, students built mathematical practices relating to ratio and proportion, which means that they internalized, organized, and reinvented a model for this mathematical content and for the learning of that content in a social environment (Gravemeijer et al., 2003). In each activity, distinct tasks encouraged students to find efficient strategies by listening to their peers, asking questions in whole-class discussions, arguing with other students, and the like. The classroom environment made students feel the need to explain their reasoning, ask questions when they did not understand critique and understand the reason of others, and use mistakes as sites for learning opportunities (Stephan et al., 2015).

METHODOLOGY

The present study aimed to find the answer to the research question with the help of classroom design research methodology using a classroom teaching experiment (Stephan, 2014). In this section, first, we briefly describe the classroom design research methodology and our rationale for using it in this study. Second, the participants of the research and the data collection tools are explained. Third, the data collection process is described for the reader to give background information about the implementation of the study. Fourth, we present the process through which we analyzed the data.

Research Design

The current study is a part of the dissertation study examining mathematical practices of seventh-grade students related to ratio and proportion concepts within the scope of design-based research methodology. Classroom design research (CDR; Stephan, 2014) was the methodology to investigate this research problem in terms of testing, analyzing, and revising both the preparation for teaching and the learning aspects of *the design*. In general, educational design research is a methodology that involves "educational ideas for student or teacher learning [that] are formulated in the design but can be adjusted during the empirical testing of the ideas." (Bakker, 2018, p. 5).

While education design research (EDR) can be conducted in various settings, CDR (Stephan, 2014) specifically studies innovations that are designed for classroom contexts. Both EDR and CDR have an interventionist structure in its nature by adjusting the research plan before, during, and after the study conducted (Plomp, 2013). Not only the instructional materials but also the proposed learning process is realized in the real classroom setting, and potential learning pathways are fostered to emerge in the design research (Cobb et al,, 2003).

CDR comprises iterative phases of design, implementation, and analysis of the innovation so that revisions can be made to strengthen the instructional sequence. This constitutes one phase of CDR; ideally, another phase takes place to implement, analyze, and revise the new instructional sequence. Macrocycles like this continue until a viable instructional sequence is ready for adaptation by teachers.

During the implementation portion of the phase, CDR utilizes a classroom teaching experiment (Cobb et al., 2003; Rasmussen & Stephan, 2008) in which a teacher implements the innovation in her classroom while researchers systematically collect data for future analysis. In our case, a classroom teaching experiment was conducted with two seventh-grade classrooms of an elementary mathematics teacher in a public school in Turkey using the ratio and proportion instructional sequence. As the classroom design research progressed, daily evaluations of the teaching and learning occurred and fed back immediately into the ongoing instruction.

Participants

The study was conducted in a public school in a major city of Turkey and was in a district with a low socioeconomic status. There were 33 students per classroom as an average. This was quite a large class size based on the OECD report about average class size in Turkey, which reported 26 students per class in public, lower secondary schools in 2017 (OECD, 2020). Two seventh grade classrooms (7/X, and 7/Y as a pseudonym, average age 12) and their mathematics teacher (Teacher Merve, as a pseudonym) participated in the study voluntarily during one semester.

We studied with two classes, 7/X (including 20 boys and 18 girls) and 7/Y (including 22 boys and 16 girls), which were overcrowded compared to the other classrooms in the school. The demographic information of the classrooms is presented in Table 6.1. The scores given in Table 6.1 represented only the average mathematics scores of the classrooms measured before the study started. Each discipline was graded out of 100, twice in an academic year. The physical setting of the classroom included one smartboard integrated with two whiteboards, one cupboard and desks.

The classrooms were inclusive classrooms in which students with learning differences learn together. There were two students with disabilities in 7/X, one student in 7/Y. Voluntary participation of the teacher, students, and their parents and awareness of the teacher about the innovations in the field of mathematics education were the reasons for preferring to work with this group. The utility of the instructional design of ratio and proportion in such a classroom is significant in understanding the variety of emerging mathematical ideas in diverse classes.

Table 6.1

Demographic Information of the Two Classrooms

	7/X	7/Y
Number of the students	38	38
Boys	20	22
Girls	18	16
Academic Achievement in Mathematics	72.63/100	72.40/100

Although the study was conducted in both classrooms, we focused on only one of the classrooms, 7/X, due to space constraints. The design team consisted of Teacher Merve, the researcher, and two field experts. Teacher Merve and the researcher actively participated in the classroom sessions. Two of them were involved in the daily evaluation of the design. Moreover, two field experts were informed about the progress of the activities and evaluated the changes, and the current process, frequently.

Data Collection Instruments

There were several sources of data collection throughout the classroom teaching experiment. The ratio and proportion instructional sequence was designed by considering relevant research based on curricular issues, learning theories, content specific theories, and the instructional design theory of RME. The study lasted approximately seven weeks for the two classrooms. During instruction, each classroom session was videotaped by two cameras, both in front of and in the back of the classroom.

Student activity papers were also collected in the process. Apart from videotaping classroom sessions and student activity sheets, the researcher took field notes every day she attended the class with the teacher. In the notes, she recorded students' learning activities, participation, contributions, or unexpected activities in the classroom. There were also several small talks with the teacher during and after the classroom sessions about the classroom progress, which were also recorded in the field notes.

Data Collection Process

Before starting the implementation, the researcher observed teacher-student interactions both in the current class and in other classes. The ratio and proportion instructional sequence was applied in 7/X, approximately 35 hours, and the activity sheets of the students had been collected and investigated. There were 23 activity sheets obtained for each student throughout the implementation. Students were asked to fill the activity sheets. These sheets were introduced like the notebook in which they may do whatever necessary changes they needed.

The duration of the activity was determined based on the performances of the students. The researcher attended in all the classroom sessions as a participant-observer. Throughout the implementation process, the researcher and the teacher focused on classroom progress. They guided the classroom with questioning, selecting different students to present, and queuing the answers according to the contribution they may make in the discussion sessions of each activity.

In the discussion environment, the teacher invited the volunteer students to the whiteboard where other peers can see their solution strategies and determine their correctness after each solution and/or idea, as suggested by Stephan and colleagues (2016). It is generally expected that students make a joint decision, but the teacher has an important role to play if everyone accepts the wrong idea or if no idea comes from the students. Therefore, the discussion phase is critical to make the emergent ideas come to the surface and to have the students present them.

Data Analysis

The data gathered from the classroom sessions were transcribed at first. These transcripts were supplemented with the insights related to the emerging ideas of the students, their learning, representations, and their way of discussing. Collected student activity sheets were also used alongside the transcripts to ascertain the progress of a specific student in the classroom. The data gathered from the discussion session in the class was analyzed using content-analysis from the transcriptions of class sessions and the artifacts of students through MAXQDA ® Plus 2020.

Qualitative content analysis is "a research technique for making replicable and valid inferences from texts (or other meaningful matter) to the context of their use" (Krippendorff, 2004, p. 18). Based on this, components of the content analysis were used: *unitizing, sampling, coding, reducing, inferring,* and *narrating,* and these components need not be a linear order (Krippendorff, 2004). For the *unitizing,* transcriptions were divided into episodes in sequential order, first. Then, episodes were *coded* under main categories such as iterating composite units, using four operations, using ratio tables, and subthemes under main themes such as drawing the units and using numeracy for units under iterating composite units.

These subcategories were selected according to the emerging ideas about patterns and recoded based on their relevancy to each other, which is called *sampling* that involves the procedure of choosing a representative sample of the episodes from the whole episodes (Krippendorff, 2004). In order to reduce, subcategories were supported by various episodes from the discussions in 23 activities. Still, episodes within the first four activities were found representative of the emerging and developed ideas about patterns for this study since the transfer of the collective knowledge to the other activities revealed the repetition of episodes in different contexts.

These key episodes were combined into detailed descriptions in the findings for the *inferring* and *narrating* part of the content analysis through field notes and student activity sheets. To understand the development of the emergent idea of an individual student, the explanations of the same students were selected, among other similar episodes.

FINDINGS

Based on the research question, students' emerging ideas were organized chronologically, providing critical episodes from the four activities of the classroom experiment. The focus was gathered around three categories: repeating patterns as a problem-solving strategy, growing number patterns with addition, and growing number patterns with multiplication.

Repeating Pattern as a Problem-Solving Strategy

The ratio and proportion instructional sequence was designed to allow students to explore proportional situations within a realistic context (e.g., aliens and food) at first and move progressively towards more abstract proportional reasoning. In the first three activities, students were never asked to use the ratio table but encouraged to develop one. In these three activities, the problems were posed so that students naturally wanted to organize the quantities in some structured way. Students created different *inscriptions* as they solved problems near the beginning of the instructional sequence. These tasks involve a pictorial representation of the rule and several problems based on this rule, as seen in Figure 6.1.

Figure 6.1

The Rule of the Ratio Given in the First Activity X(b), Third Question in the First Activity X(b)

(a) RULE:

Explanation: 1 food bar feeds three aliens.

(b) How many food bars are needed to feed 15 aliens?

Source: Stephan et al. (2015).

Most students used division to solve this problem; however, merely doing a calculation was not enough; they were expected to explain why their calculation worked for the given situation. When it came to explanation, students had difficulty explaining the meaning of the dividend, divisor, remainder, and the quotient for the characters in the problem. For example, if a student computed $15 \div 3 = 5$, they had difficulty explaining that the operation 15 divided by 3 meant that they were organizing the 15 aliens into groups of 3 aliens (also equivalent to 1 food bar) to get 5 groups of 3 aliens and 5 groups of food bars.

Several students who used the division algorithm and were unable to provide a satisfying explanation changed their strategy by drawing repeating quantities, that is, creating and reasoning with patterns. Egemen was a student, for example, who explained it on the board to his friends and received approval in the whole-class discussion (see Figure 6.2). Egemen solved the third problem in the first activity X(b) in his mind through the division algorithm, but when he came to the board, he directly drew the rule and repeated that rule by drawing five times until he got 15 aliens in total. Drawing the initial ratio and repeating it helped Egemen to explain his result.

> Egemen: Hocam (Ma'am or Teacher), since the rule is "one food bar feeds three aliens," I drew it like that (while drawing on the board).

Teacher Merve: Okay, Egemen! Can you explain how you found the result?

Egemen: The rule is "one food bar feeds three aliens." Every three aliens eats one food bar (Drawing ellipses around each of three aliens on the board to show five groups of three); therefore, the result is five food bars needed in total.

Figure 6.2

Egemen's Explanation for the Third Problem in the First Activity

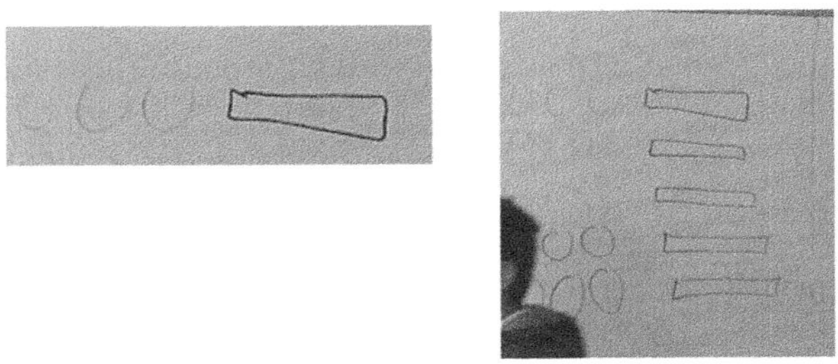

Egemen provided one of the typical solution strategies: "to identify the rule of the ratio for the problem and model it" and "to repeat that model by drawing" and "find the missing value." Students eventually determined that drawing out all aliens and food bars was not a time-efficient solution for ratio and proportion problems. Drawing each value on the board required extra time, especially in problems that purposely included high numbers such as 72 aliens.

Therefore, the learning environment and instructional activities provoked students to develop more efficient solution strategies for the next time. In the second activity, the rule of the ratio was "one food bar feeds 3 aliens" again, and the problem asked, "Will 12 food bars be enough for 36 aliens? Explain." This time, the numerical values given in the activities were increased, and the comparison problem, instead of a missing value problem, was used to provoke students to develop different strategies.

Students were efficiently using patterns to find a logical solution. At this time, one of the students, Burcu, provided a unique solution practiced by other students in the next problems. When she said that she used the drawing on the board, her peers thought that she would use the strategy of Egemen. They objected to drawing 36 aliens one by one with good reason.

Burcu: Teacher! Look at here, please! When we do all this three by three, here is a total of 36. That's enough [draws 12 rectangles to stand for 12 food bars and places the numeral 3 in each (see Figure 6.3).

Figure 6.3

Burcu's Drawing for Solving the Question, Will 12 Food Bars Be Enough to Feed 36 Aliens?

Teacher Merve: I see! You did it like that, "I did not draw the aliens, and I wrote the numbers in it (food bars)" so when you said 3, 3, 3, it was 36 in total. Okay, nice.

Instead of drawing aliens, Burcu drew the food bars wrote the number value of the aliens in the rule inside each rectangle. She modeled the rule at first with one rectangle instead of two different pictures (circle and rectangle) and repeated that model 12 times to test whether 12 food bars were enough for 36 aliens. She set aside the 36 aliens mentally, and she tried to find the number of aliens as the missing value.

There were 12 food bars drawn on the board, and she compared this result with the problem, multiplied it by 3, and concluded that "That's enough." Her way of repeating patterns led her to multiply and also explain her reasoning. This solution was not rejected anymore, and again, several students used this solution strategy in the next problems, as provided in Figure 6.4.

In the third activity, tasks continued with a rule "one food bar feeds 5 aliens." The question asked, "How many aliens can you feed with five food bars? Explain. Students were constructing new strategies by using multiplication and division algorithms and developing their justifications each time. Simultaneously, several students continued using the pictorial repetition of the rule but evolved it each time.

One of the students, Can, transformed Burcu's strategy into a *more symbolic way,* as given in Figure 6.5. This time, he neither drew the food bar nor drew the aliens; he just wrote the number values of each unit. In his model, he used the equation sign to express "feeding" action. "Five aliens eat one food bar" rule was represented number values written in a square and Can repeated this rule five times to express 5 groups of food bar-alien ratios.

Figure 6.4

Other Students' Practicing Burcu's Strategy

Figure 6.5

Can's Remodel of Burcu's Strategy for the Next Task

Larger numbers became a source of motivation for the students to develop their models or adopt new models to reduce the time amount for a solution as it was revealed with the inscriptions of Burcu and Can. Moreover, they did not leave iterating the unit by drawing repeating patterns since drawing was mentioned as a "more concrete solution" at first, and a few students presented multiplicative understanding with justifiable reasoning. To iterate the unit precisely and see them made the students feel safe with their justifications for the solution. Using repeating patterns by drawing was not enough to achieve multiplicative thinking.

The learning environment was designed to reinforce students' thinking towards proportional reasoning with the new challenges in the following activities. Therefore, these examples were left for further discussions to be the start-ups for the ratio table as provided below.

Growing Number Patterns With Addition

In further activities, students found new ways to organize their ideas as prompted by both a change in instructional activities and classroom discussion. The problem of the previous students' strategies was that each representation had its own language, and each time students added their remark; it made students a little bit confused. In essence, they were using principles of ratio tables but with different representational approaches.

Being aware of this situation, Teacher Merve and the researcher encouraged students to make a table using the data given in the problem, which made the number values more organized. The third activity was also designed to provoke students to reason with a table. The rule of the third activity was "One food bar feeds 5 aliens," and the third problem stated, "Using a table, show how many food bars you would need to feed 70 aliens." One of the students, Deren, represented her answer on the board. Before her representation, how to draw a table was discussed in the classroom.

Specifically, the students first had to decide how many unique quantities to include in a table (i.e., aliens and food bars) and then debated different ways to organize the quantities in a table (e.g., deciding whether to write these names in the table). Students did not ask about how to write numbers on the table. Each student had their own organization in their mind. Deren created a table on the board, as given in Figure 6.6.

Figure 6.6

Deren's Long Table Representation

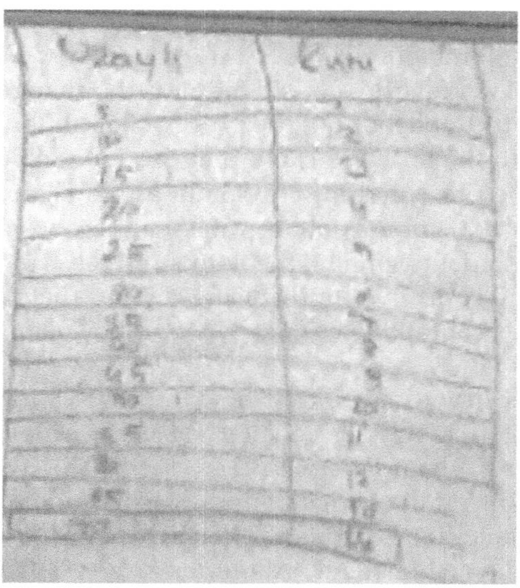

Alien	Food Bar
5	1
10	2
15	3
20	4
25	5
30	6
35	7
40	8
45	9
50	10
55	11
60	12
65	13
70	14

Deren created her table extended vertically because, being the tallest person in the classroom, she could easily use the top of the whiteboard, which could be seen from everywhere in the classroom. Since the shape of the space given in the activity sheet and the whiteboard was rectangular (length is way bigger than width), a horizontal table was more practical to extend to the right of the table. This change in the form of the ratio table created by the physical conditions for writing did not influence students' mathematical ideas. We could see horizontal ratio tables in further examples.

As in Deren's example, two columns showed the names of the values: aliens and food bars. Her reason to start with the "alien" column originated from the problem with 70 aliens and asking the missing value food bars. She began writing the numbers under the "alien" column, and then she wrote the numbers under the "food bar" column. Under those names, the numbers were placed in increasing order. The "alien" column started with 5 and increased by 5 in the column. In the other column, "food bar," she increased one by one starting from 1. The increase in the numbers was decided by the rule given in the problem, and she developed two different growing number patterns in the table.

After repeating the numbers, this was an evolution of thinking in the classroom to use growing patterns. When they used repeating patterns in previous problems, they also added the numbers in each step, saying it verbally, but this time pattern was more explicitly visible for the discussion. She used a long build-up ratio table by using addition with constant number values. They were not iterating the model of the units, but this time they were iterating the constant difference. Burcu's strategy did not help students see the result of the problem directly. Each time, they were adding the numbers together and writing the result. This time, the change was that the result was written in the table but not the process of addition.

Another student, Ferit, came to the board to claim that this long table was not satisfying for him, and he presented his *new* solution (see Figure 6.7). A first question came from the students about whether there were enough fives in the cells, and two students counted the number of cells, and then Ferit counted them. There were 14 fives in the cells. He said that "I added five[5]s and then, teacher, I counted five[5]s and here is how many [foodbars] were needed."

This representation brought a new discussion of how to draw a ratio table and its properties. After the whole class agreed that *there were names of the unit on the table, there should be an answer to the problem*, Alp asked, "How do we know the question just by looking at the Ferit's solution?" The students tried to make Ferit's solution a "table" but could not. Then Teacher Merve asked, "Does Deren's solution make a table, then?" All students agreed that it was a table, which was suitable based on their explanations about the properties of the table.

Teacher Merve also probed students to consider another difference between the two representations. Ferit used "5" as the number of aliens in the rule, and Deren used "5" as the constant difference between the aliens

Figure 6.7

Ferit's New Solution

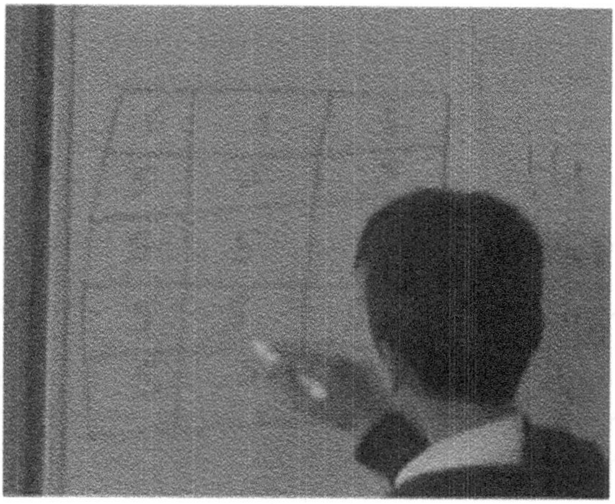

in two consecutive rows and additionally "1" for the constant difference between the food bars in two consecutive rows. In this way, Teacher Merve called students' attention to another pattern within the table.

Growing Number Patterns With Multiplication

The way students used the ratio table supported their use of repeated addition and multiplication as well. To reinforce students' multiplicative thinking, the researcher and Teacher Merve used the term "short table" by highlighting how Deren and Ferit spent lots of time to solve this problem with addition. In the growing patterns with the addition given above, each column was handled in itself, still not dependent on each other. While using long build-up strategy, the focus was on the one column at a time, the correctness of the pattern, the increase between the rows in one column: 5,10,15 . . . 70 and 1, 2, 3, ... 14.

However, there was also another relationship between two columns in the same row. This situation was enlightened by the students' different solutions in the whole-class discussion.

> Teacher Merve: Look, Deren wrote aliens and food bars (showing the columns), then she found five, five aliens first and then found the number of food bars in return, is it true, Deren?

Deren: Yess!

Researcher & Teacher Merve: Is there anyone who did in a short way?

Researcher: Cansu, can you show us your explanation? [In the Explore Phase of the Activity, the researcher saw Cansu's short form of explanation]

Cansu came to the board and drew a table (see Figure 6.8).

Researcher: Cansu, can you explain to us how you found it?

Cansu: Ma'am/Teacher, one food bar feeds 5 aliens. This is 5 times the number of food bars. Therefore, 70 (aliens) will be 5 times the number of food bars. Teacher, if 5 aliens are fed with one food bar, 70 aliens then ... (calculates, divides 70 into 5 and writes the result 14 in the empty cell in the table)

Deren: This is very short, mine is long.

In the Explore Phase of LED, the researcher had seen Cansu's explanation involving a short table for the problem. The researcher and Teacher Merve decided to call her to the whiteboard in order to introduce a short table to her peers after Deren's table. The advantage of the short table is that not only is it a more efficient way to represent repetitive iterations of the ratio, but it can also begin to support multiplicative relationships within a ratio and between equivalent ratios. In other words, the short table can better support multiplicative interpretations of the patterns that are represented in the table.

Cansu drew a vertical table and used "alien" and "food bar" as the names for the columns, similar to Deren. She explained that she put the values in the first row according to the rule: 1 for the food bar, 5 for the aliens. She explored the relationship between the two cells/units in the same row, 5 times (see Figure 6.8). The constant relationship between the cells in the same row was the ratio between two units: food bars to aliens. She expressed this relationship with an arrow from left to right, as shown in Figure 6.8.

The direction of the arrow was meaningful as well because it meant division with 5 if the arrow was directed from right to left as given in the second row in Figure 6.8. What she did was to divide 70 aliens into 5, and she found 14 as a result. All in all, this short table was tiny, time-efficient, and represented a multiplication relationship as suggested, but the classroom needed some time and practice to internalize this process.

Figure 6.8

Cansu's Short Table Representation

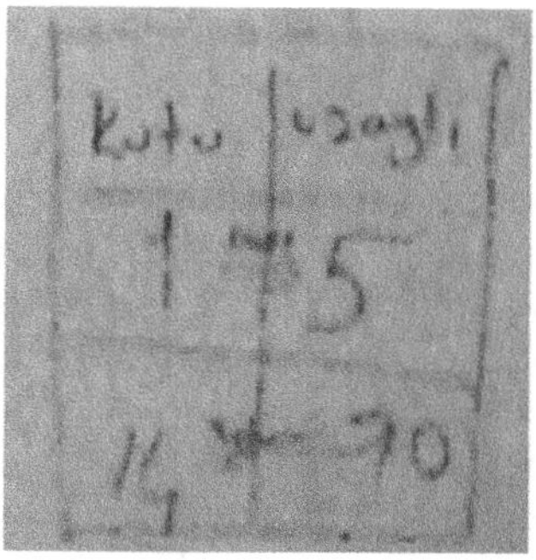

Food Bar		Alien
1	Multiplied by 5 →	5
14	← Divided by 5	70

While internalizing, students started using multiplication as a more efficient operation between two units.

In activity four, the rule was "two food bars feed 4 aliens." It was the first time that the rule was not a unit ratio. The third problem was, "How many aliens can 14 food bars feed? Explain." In the previous activities, it was observed that Burcu used repeating patterns. Still, after the whole-class discussion in the class, Burcu evolved his solution by representing the multiplicative relationship between the two units, as given in Figure 6.9.

Figure 6.9

Burcu's Representation of the Relationship Between Two Units. (ACTIVITY SHEET A4)

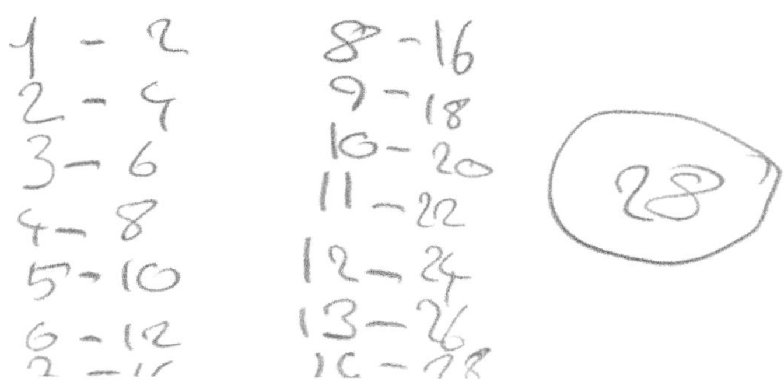

First, Burcu wrote the name food bar since the number of food bars was given in the problem, and the number of aliens was the missing value.

Burcu listened to the previous discussion about finding the unit ratio and used the multiplicative relationship as "one food bar feeds 2 aliens." The number written in the left part was the number of food bars, and the number written in the remaining part was the number of aliens. The line between the cells represented the relationship between the numbers. She started from 1 food bar to 2 food bars. The number of aliens was two times greater than the number of food bars.

Although she did not draw any arrow kind of symbol to represent the relationship between alien and food bar, she followed an order from left to right "one food bar feeds two aliens, two food bars feed four aliens." Using this calculation, she replicated this rule of the pattern until she reached number 14 in the food bar part. She utilized the idea that two is a constant multiplier between the numbers, not independent units this time. As it was remembered from Deren's explanation in Figure 6.5, she used considerable time and space on the whiteboard to draw the table in that length.

Teacher Merve and the researcher asked the students, "Is there any other way to save time to solve this problem?" in order to prepare them for a subsequent problem, "How many aliens will 98 food bars feed? Explain." This time the number of aliens was the largest among all, and the hope was that students would curtail drawing the aliens and food bars in some quicker manner. Since the multiplication strategy was still new to the students, when Cansu used it again (see Figure 6.10), they were not used to it.

Figure 6.10

Cansu's Explanation of the (TRANSCRIPT C4, Pos. 253)

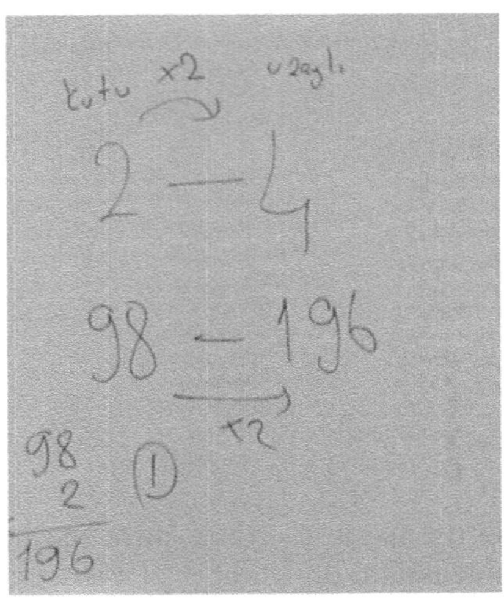

The goal of the fourth activity was to support students in curtailing their additive patterns and use a short table. Only a few students created a short table, so within the whole-class discussion, Teacher Merve continued asking questions:

> Teacher Merve: Well, I'll say something. Suppose you took an exam, you had a limited amount of time. How would you do that? Were you dealing with ratio table, dealing with drawing shapes, how would you do it? Ercan?
>
> Ercan: I would use division and multiplication directly as well. But I used the table this time.
>
> Teacher Merve: You would do division and multiplication! Okay, you are saying this is more practical. So, is it necessary to create a long table after understanding its efficient way? Making a table may provide a more in-depth explanation to understand the problem, yes, but no need for writing all numbers; for example, we can make the table short.
>
> Several students in the classroom: I did it like this! (see Figure 6.11).
>
> Teacher Merve: Can we put dots in between and make the table short? Well, I'll show you something.... Guys, take a look here. Food bars, number of aliens.... Suppose I will make a ratio table. How many aliens can be fed with one food bar?

Teacher Merve was filling the gap between Burcu's and Deren's solution strategy. With her explanation, she made it more concrete to help students create a connection between building up with addition and shortening with multiplication. She replaced the number of food bars in the first-row one by one and then filled the second row with the number of aliens by multiplying the number of food bars with two, as given in Figure 6.11.

Figure 6.11

Teacher Merve's Representation of Short-Table

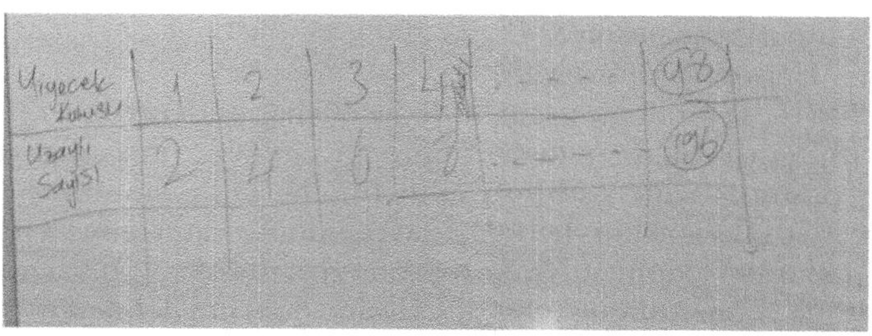

She filled the cells by question-answer with the whole class. By emphasizing this constant relationship among two units, food bars and aliens, on the dots and 98 food bars, students automatically multiplied 98 by 2 and found a result of 196. This was one of the relationships within the table between two distinct units. What about the relationship between the two same units?

In the other problems of the fourth activity, another constant relationship emerged, and it became another strategy for solving the problems. Two students brought up this issue in the discussion, as given in Figure 6.12. This time they found a pattern among the same units. If the multiplicative relationship between the two cells of the food bars was 7, then this relationship between the number of aliens is the same as food bars, 7. This constancy keeps throughout the table, and it shows the variety of relationships among the variables given in the tables.

Figure 6.12

Representations of the Multiplicative Relationship Between Two Same Units

DISCUSSION

In the current study, seventh-grade students created different strategies to accomplish the learning goals identified for each activity during the whole-class discussion. While students were developing an understanding with ratio tables, they used several patterning strategies. Mathematical reasoning while creating ratio table fostered students to use the successive manipulation of numbers to maintain an invariant relationship between the two quantities in the ratio; therefore, exploring patterns within ratio tables can be strong support for proportional reasoning as mentioned in the literature (Abrahamson & Cigan 2003; Ercole et al., 2011; Middleton & van den Heuvel-Panhuizen, 1995).

Based on these findings, it can be deduced that students' emerging ideas began with repeating the pattern within a ratio table, moved to interpret growing patterns with addition, and finally, understanding constant multiplicative relationships between the cells in the ratio table. These ideas represented a development from repeating patterns to growing patterns with multiplication. Although this description does not define a hierarchical pathway for the students, it describes the naturally occurring reasoning that emerged in classroom discussions.

Exploring the emerging ideas about patterns through this development process provided a relationship with the development of multiplicative reasoning of the students as well. Similar to Warren and Cooper (2007), we argue that repeated patterning can serve as a means to introduce the ratio concept. Additionally, with the help of repeating patterns, Warren and Cooper claimed that students could easily identify ratios and equivalence between ratios, and utilize an abstract notation system.

Similarly, Battista and van Auken Borrow (1995) contended that iterating composite units serves as a basis for later multiplicative reasoning. We found evidence of this in that our students iterated composite units (aliens and food bars) in drawings and more efficiently in build-up tables and short ratios tables. In the current study, we showed similar findings and provided an example of an inquiry-based learning environment which involves ratio table tool and students'developing and working on it through patterning.

Growing number patterns with addition is a building up strategy in which students found a context-specific ratio and then iterated that ratio to build up additively to the missing value (Steinthorsdottir, 2006). Then, they explored the meaning of the additive patterns within the ratio table or drawing. This reasoning was a helpful transition step but needed to be developed into multiplicative thinking (Lamon, 2012; Steinthorsdottir, 2006) to fully understand proportional reasoning. As Lamon (2012) asked, "Do they focus on the relationship between the two quantities?" (p. 111).

The instructional sequence was designed to bring out whether students see the multiplicative relations while using a build-up strategy. Besides, Teacher Merve understood the importance of this learning goal and emphasized multiplicative interpretations of the patterns in a whole-class discussion. In this study, Teacher Merve used the natural transition from drawing figures to writing symbols as a starting point for these discussions. Since students did not naturally interpret the patterns in multiplicative ways, Teacher Merve prompted students to find more efficient ways to represent the growing pattern.

Through this guidance and purposive discussions, the patterns in the build-up strategy provided a smooth transition from addition to the multiplication and finally a multiplicative interpretation of ratios and equivalent ratios. This kind of transition through growing number patterns with addition may encourage multiplicative reasoning in ratio.

Using patterns had become part of ratio table development, and the conversations around the ratio table also helped students improve their patterning reasoning within an inquiry-based environment. Moreover, it has been found that middle-grade students perform better while solving ratio problems within a learning environment, which reinforces the utility of strategies to reason about multiplicative relationships (Ben-Chaim et al., 1998; Cramer & Post, 1993).

Our findings are consistent with Warren and Cooper's (2007) study, which identified several strategies of the students while using the table. They explained that students developed some "down" and "across" strategies within the table, similar to Deren's explanation of growing number patterns with addition.

Additionally, students used illustrations for the multiplicative relationship between different and similar units within a growing number pattern. Investigating the unit of repeat and iterating that unit were bridges between ratios and the patterns that their iterations form (Papic, 2007; Papic et al., 2011; Warren & Cooper, 2008). Although the grade level of the students may differ, students may come across similar challenges to overcome, and they similarly build their knowledge to develop multiplicative structures.

CONCLUSION

As a result, our contributes to research in that we found exploring patterns illustrated in a ratio table (both long and short) can lead to students' development of multiplicative structures. We have provided a set of instructional materials and the outline of a reinvention process with those materials and discourse that can support multiplicative interpretation of ratios.

This study was limited to the classroom teaching experiment designed for seventh-grade students within the context of ratio and proportion instructional sequence. The activities were not intended to teach patterns, but the students explored ways of incorporating patterns to deal with the challenge given in the problems. Without going out of the context, pattern related discourse can be embedded in the instructional sequence to enrich the classroom environment by making connections between ratio and pattern contexts.

For example, in other classroom experiments with this instructional sequence, we have seen students notice that as the numbers in the ratio table increase additively, the difference between them vertically does as well. For example, for the table 1:5, 2:10, 3:15, 4:20, and so forth, the difference between the numbers in the first ratio is 4, the difference in the second is 8, and continues with 12 and 16. This is an additional pattern that can be explored and explained by students as well. New implementation of the ratio instructional sequence will capitalize on other patterns that can be noticed within ratio tables to strengthen multiplicative reasoning.

REFERENCES

Abrahamson, D., & Cigan, C. (2003). A design for ratio and proportion instruction. *Mathematics Teaching in the Middle School, 8*(9), 493–501.

Bakker, A. (2018). *Design research in education: A practical guide for early career researchers*. Routledge.

Battista, M. T., & van Auken Borrow, C. (1995, October). *A proposed constructive itinerary from iterating composite units to ratio and proportion concepts* [Paper presentation]. The annual meeting of the North American Chapter of the International Group for Psychology of Mathematics Education, Columbus, OH.

Billstein, R., Libeskind, S., & Lott, J. (2013). *A problem-solving approach to mathematics for elementary school teachers* (12th ed.). Pearson Higher Education.

Behr, M. J., Harel, G., Post, T. R., & Lesh, R. (1992). Rational number, ratio, and proportion. In D. A. Grouws (Ed.), *Handbook of research on mathematics teaching and learning: A project of the National Council of Teachers of Mathematics* (pp. 296–333). Macmillan.

Ben-Chaim, D., Fey, J. T., Fitzgerald, W. M., Benedetto, C., & Miller, J. (1998). Proportional reasoning among 7th grade students with different curricular experiences. *Educational Studies in Mathematics, 36*(3), 247–273. https://doi.org/10.1023/A:1003235712092

Broekman, H., van der Valk, T., & Wijers, M. (2000, August). *Teacher knowledge needed to teach ratio and proportion in secondary school mathematics* [Paper presentation]. The 25th ATEE Conference, Barcelona, Spain.

Burns, M. (2015). *About teaching mathematics: A-K 8 research* (4th ed.). Math Solutions.

CCSSO (Council of Chief State School Officers). (2010). *Common core state standards*. http://corestandards.org

Cobb, P., Confrey, J., DiSessa, A., Lehrer, R., & Schauble, L. (2003). Design experiments in educational research. *Educational Researcher, 32*(1), 9–13. https://doi.org/10.3102/0013189X032001009

Cramer, K., & Post, T. (1993). Making connections: A case for proportionality. *Arithmetic Teacher, 60*(6), 342–346. http://www.jstor.org/stable/41195594

Dole, S. (2008). Ratio tables to promote proportional reasonings in the primary classroom. *Australian Primary Mathematics Classroom, 13*(2), 18–22.

Ercole, L. K., Frantz, M., & Ashline, G. (2011). Multiple ways to solve proportions. *Mathematics Teaching in the Middle School, 16*(8), 482–490.

Gravemeijer, K., Bowers, J., & Stephan, M. (2003). Chapter 4: A hypothetical learning trajectory on measurement and flexible arithmetic. *Journal for Research in Mathematics Education Monograph, 12*, 51–66.

Howe, C., Nunes, T., & Bryant, P. (2011). Rational number and proportional reasoning: Using intensive quantities to promote achievement in mathematics and science. *International Journal of Science and Mathematics Education, 9*(2), 391–417. https://doi.org/10.1007/s10763-010-9249-9

Inhelder, B., & Piaget, J. (1958). *The growth of logical thinking from childhood to adolescence.* Basic Books.

Kaput, J. (1994). The representational roles of technology in connecting mathematics with authentic experience. In R. Biehler, R. W. Scholz,

R. StraBer, & B. Winkelman (Eds.), *Didactics of mathematics as a scientific discipline* (pp. 379–397). Kluwer.

Krippendorff, K. (2004). *Content analysis: An introduction to its methodology* (2nd ed.). SAGE.

Lachance, A., & Confrey, J. (2002). Helping students build a path of understanding from ratio and proportion to decimal notation. *The Journal of Mathematical Behavior, 20*(4), 503–526. https://doi.org/10.1016/S0732-3123(02)00087-1

Lamon, S. J. (2012). *Teaching fractions and ratios for understanding: Essential content knowledge and instructional strategies for teachers* (3rd ed.). Routledge.

Markworth, K. A. (2012). Growing patterns: Seeing beyond counting. *Teaching Children's Mathematics, 19*(4), 254–262. https://doi.org/10.5951/teacchilmath.19.4.0254

Middleton, J. A., & van den Heuvel-Panhuizen, M. (1995). The ratio table. *Mathematics Teaching in the Middle School, 1*(4), 282–288.

Mulligan, J., & Mitchelmore, M. (2009). Awareness of pattern and structure in early mathematical development. *Mathematics Education Research Journal, 21*, 33–49. https://doi.org/10.1007/BF03217544

Organization for Economic Co-Operation and Development (OECD). (2020). *Average class size.* https://stats.oecd.org/Index.aspx?DataSetCode=EDU_CLASS

Papic, M. (2007). Promoting repeating patterns with young children-More than just alternating colours!. *Australian Primary Mathematics Classroom, 12*(3), 8–13.

Papic, M., & Mulligan, J. T. (2007). The growth of early mathematical patterning: An intervention study. In J. Watson & K. Beswick (Eds.), *Mathematics: Essential research, essential practice*. Proceedings of the 30th annual conference of the Mathematics http://citeseerx.ist.psu.edu/viewdoc/download?doi=10.1.1.842.9776&rep=rep1&type=pdf#page=594

Papic, M., Mulligan, J. T., & Mitchelmore. M. C. (2011). Assessing the development of preschoolers' mathematical patterning. *Journal for Research in Mathematics Education, 42*(3), 237–269. doi:10.5951/jresematheduc.42.3.0237

Plomp, T. (2013). Educational design research: An introduction. In T. Plomp & N. Nieveen, (Eds.) *Educational design research-An introduction: Part A* (pp. 5–9). Netherland Institute For Curricuum Development (SLO).

Rasmussen, C., & Stephan, M. (2008). A methodology for documenting collective activity. In A. E. Kelly, R. A. Lesh, & J. Y. Baek (Eds.), *Handbook of innovative design research in science, technology, engineering, mathematics (STEM) education* (pp. 195–215). Taylor and Francis.

Resnik, M. (1982). Mathematics as a science of patterns: Epistemology. *Noûs, 16*(1), 95–105. https://doi.org/10.2307/2215419

Resnik, M. D. (2005). *Mathematics as a science of patterns*. Oxford University Press.

Singh, P. (2000). Understanding the concepts of proportion and ratio constructed by two grade six students. *Educational Studies in Mathematics, 43*(3), 271–292. https://doi.org/10.1023/A:1011976904850

Sozen Ozdogan, S., Akyuz, D., & Stephan, M. (2019). Developing ratio tables to explore ratios. *Australian Mathematics Education Journal, 1*(2), 16–21.

Steffe, L. P. (1994). Children's multiplying schemes. In G. Harel & J. Confrey (Eds.), *The development of multiplicative reasoning in the learning of mathematics* (pp. 3–39). SUNY Press.

Steinthorsdottir, O. B. (2006, July). Proportional reasoning: Variable influencing the problems difficulty level and one's use of problem solving strategies. In Novotná, J., Moraová, H., Krátká, M. & Stehlíková, N. (Eds.), *Proceedings 30th Conference of the International Group for the Psychology of Mathematics Education, Vol. 5, pp. 169–176.* PME.

Stephan, M. (2014). Conducting classroom design research with teachers. *ZDM, The International Journal of Mathematics Education, 47*(6), 905–917. https://doi.org/10.1007/s11858-014-0651-6.

Stephan, M., McManus, G., Smith, J., & Dickey, A. (2015). *Ratio and rates.* https://cstem.uncc.edu/sites/cstem.uncc.edu/files/media/Ratio%20T%20Manual.pdf

Stephan, M., Pugalee, D., Cline, J., & Cline, C. (2016). *Lesson imaging in math and science: Anticipating student ideas and questions for deeper STEM learning.* Association of Supervisors and Curriculum Designers.

Tourniare, F., & Pulos, S. (1985). Proportional reasoning: A review of the literature. *Educational Studies in Mathematics, 16,* 181–204. https://doi.org/10.1007/PL00020739

Van de Walle, J. A., Karp, K. S., & Bay-Williams, J. M. (2016). *Elementary and middle school mathematics: Teaching developmentally* (9th ed.). Pearson Education.

Van den Heuvel-Panhuizen, M., & Drijvers, P. (2020). Realistic mathematics education. In S. Lerman (Ed.), *Encyclopedia of mathematics education* (pp. 713–717). Springer. https://doi.org/10.1007/978-94-007-4978-8_170

Warren, E., & Cooper, T. (2007). Repeating patterns and multiplicative thinking: Analysis of classroom interactions with 9-year-old students that support the transition from the known to the novel. *The Journal of Classroom Interaction, 41*(2), 7–17.

Warren, E., & Cooper, T. J. (2008). Generalising the pattern rule for visual growth patterns: actions that support 8 year olds' thinking. *Education Studies in Mathematics, 67,* 171–185. https://doi.org/10.1007/s10649-007-9092-2

Warren, E., Miller, J., & Cooper, T. (2012). Repeating patterns: Strategies to assist young students to generalise the mathematical structure. *Australasian Journal of Early Childhood, 37*(3), 111–120. https://doi.org/10.1177/183693911203700315

Wittmann, E. C. (2005, July). *Mathematics as the science of patterns-a guideline for developing mathematics education from early childhood to adulthood* [Paper presentation]. Plenary Lecture at International Colloquium: Mathematical learning from Early Childhood to Adulthood (pp. 1–25), Mons/Belgium. https://mathinfo.unistra.fr/websites/math-info/irem/Publications/Annales_didactique/vol_11_et_suppl/adsc11supplweb_wittmaneng.pdf

CHAPTER 7

USING WORKED EXAMPLES AS A SCALABLE PRACTICE FOR TEACHING FRACTION MAGNITUDE AND FRACTION COMPUTATION

Kathryn J. deVries, Julie L. Booth, Laura K. Young, Christina Areizaga Barbieri, Emunah Mager Garfield, and Kristie J. Newton

INTRODUCTION

Recent findings provide empirical evidence for the long-held belief that having a sound foundational knowledge of fractions is critical for students learning algebra (e.g., Barbieri et al., in press). This is likely because fraction knowledge builds a rich number knowledge and an accuracy and fluency in a variety of rational number procedures that are then built upon in algebra courses. This study examines fraction instruction that is structured in a way to minimize cognitive load for the learner and support the development of conceptual and procedural fraction knowledge. Having students explain worked examples targets common misconceptions about fraction problem features and incorporates visual representations. We anticipate that the worked example approach will yield an appreciable improvement in students' fraction computation skills and increase understanding of fractions

Mathematics as the Science of Patterns:
Making the Invisible Visible to Students Through Teaching, pp. 127–151

as numbers on the number line alongside whole numbers. We explore how students explaining fraction magnitude in a worked example may yield additional unique benefits to fraction magnitude knowledge.

The Key to Improving Students' Mathematical Proficiency

The U.S. Department of Education issued a call to action with *Foundations for Success: The Final Report of the National Mathematics Advisory Panel* in 2008, asserting that "without substantial and sustained changes to its educational system, the United States will relinquish its leadership in the 21st century" (U.S. Department of Education, 2008, p. xi). In the last decade, math researchers and educators have sharpened their focus on helping students learn to reason with mathematics rather than follow a recipe of steps, and algebra knowledge in particular has been identified as critical for mathematical proficiency (Siegler et al., 2012). Successful learning in algebra is linked to greater success in higher mathematics (Lee & Mao, 2020) and career outcomes (Wu, 2001). Researchers looking to improve students' learning of critical algebraic knowledge have identified a major factor that may open the door to success in algebra and beyond: a sound foundational knowledge of fractions (Brown & Quinn, 2006; Rodrigues et al., 2016; Siegler et al., 2012; Wu, 2001).

One of the reasons that fraction knowledge is thought to prepare students to learn algebra is that successfully solving algebraic equations relies on the ability to apply a variety of rational number procedures both accurately and fluently (Brown & Quinn, 2006). Another possible reason is that the rich number knowledge that results from a conceptual understanding of fractions supports students in managing the generalized arithmetic and abstraction required in algebra (Booth & Newton, 2012; Wu, 2001). Because of its relation to algebra and other higher-level math, it is perhaps not surprising that the teaching of fractions has become an intense focus for educators and researchers, and a critical target for study for federal research funding. For instance, the National Science Foundation (NSF, n.d.) has active and expired awards totaling over $67M for projects which use "fractions" as a keyword (https://www.nsf.gov/awardsearch/). In addition, over $111M has been awarded by the Institute of Education Sciences (IES) since 2005 for projects involving fraction (https://ies.ed.gov/funding/grantsearch/). Thus, over $178M has been spent by these two agencies alone to study students' learning about fractions and on software, curriculum, and professional development to improve students' learning about fractions. Why has this level of time and money commitment been deemed necessary to identify best practices for teaching fractions? That is, what is so complicated about learning fractions, and what do those difficulties suggest about best instructional practices?

Fraction Learning Difficulties and Instructional Implications

One key type of difficulties is specific to students' conceptual and procedural knowledge related to carrying out computations with fractions. We define procedural knowledge as knowledge of mathematical operations—the specific sequence of steps or actions to take to reach a certain goal (Rittle-Johnson & Schneider, 2015)—and conceptual understanding as knowledge of general mathematics ideas as well as principles that underlie procedures (i.e., knowing why a procedure works) as proposed by Crooks and Alibali (2014). When it comes to fraction knowledge, both the concepts and the procedures often seem contradictory to what a child may have previously learned. In particular, students attempt to overgeneralize their procedural knowledge for whole numbers to other types of numbers. For instance, when learning to multiply whole numbers in elementary school, students come to internalize rules such as *multiplication yields a larger answer* (Siegler & Pyke, 2013). While this is indeed true for whole numbers, it is not true for all numbers, and erroneously generalizing this rule to rational numbers likely causes confusion. This whole number bias may, thus, obstruct students' ability to understand key principles related to other types of numbers, including that multiplying fractions can yield smaller answers and dividing fractions can yield larger answers (Jordan et al., 2013). Unfortunately, in many mathematics domains, students attempt to solve problems without conceptual understanding of the procedure that they are using. In fact, the development of procedural knowledge sometimes precedes the development of conceptual knowledge in mathematics (Rittle-Johnson & Siegler, 1998). Helping students understand the underlying meaning of the procedures they are using should help them develop more effective conceptual knowledge and also become more proficient with the procedures (Siegler et al., 2010). Consistent with this assertion, a variety of studies have provided causal evidence that making fraction instruction meaningful improves fraction computation skill in students of varying ages (e.g., Anand & Ross, 1987; Cramer et al., 2002; Hudson Hawkins, 2008; Irwin, 2001; Mack, 2000; Nishida, 2008). One way to improve meaning-making in fraction computation is to employ external representations to show computational procedures, such as rectangles divided into a certain number of equal parts for understanding the need to find a common denominator when adding or subtracting (Siegler et al., 2010) or for thinking about how to find a part of a part, as is needed in multiplication of two fractions (Sowder et al., 1998). In addition, instruction should address common misconceptions regarding procedures with fractions (Siegler et al., 2010), as it is well established that misconceptions about problem features hinder performance and learning when it comes to problem-solving (e.g., Booth & Koedinger, 2008). A number of common misconceptions have been identified for fractions, including thinking that denominators are treated the same way during addition and multiplication (Painter, 1989), believing that any whole numbers in a problem share a denominator with the fractions in the problem (Mack, 1995), or believing that numerators and denominators are separable (Kerslake, 1986).

An additional issue is that students struggle to develop conceptual understanding of what fractions are and how they fit into the number system; students find this difficult, and even teachers have been shown to have similar difficulty with conceptual fraction knowledge (Moseley et al., 2007). Learners often fail to understand the relationship between the numerator and the denominator, sometimes even disregarding the denominator altogether (Rodrigues et al., 2016). Further, they struggle to see fractions as numbers in and of themselves, with properties such as magnitudes and the capacity to be located on a number line with whole and other types of numbers (Rodrigues et al., 2016). Using number lines to represent fractions can help students gain a sense of the fractions' relative magnitudes and their status as part of the number system (Siegler et al., 2010); number line estimation has been shown to be highly predictive of students' fraction procedural and conceptual knowledge (Jordan et al., 2013). Several studies provide causal evidence that training students to think about fractions and decimals as numbers—typically by having them represent them on number lines (Barbieri et al., 2020; Rittle-Johnson et al., 2001; Keijzer & Terwel, 2003; Schneider et al., 2009)—improves students' conceptual knowledge of fractions and other types of rational numbers. In order to clarify that fractions are numbers with quantities, students should be given opportunities to think about how fractions (and other rational numbers) fit into the number system as represented on a number line (Siegler et al., 2010). The nature of the activities and the number lines may vary, from number lines with multiple sets of labels, which can help students see the relations between two different denominators (Beckmann, 2008), to minimally labeled lines, which force students to think about how to divide up the space to accommodate different denominators (Van de Walle et al., 2010). Such activities may be difficult, but they can help students to think about a variety of aspects of the numerical values of fractions, such as ordering, relative sizes of two fractions, and how whole numbers are also fractions (Siegler et al., 2010). Because number lines are useful for supporting knowledge about fraction magnitudes, they can also effectively demonstrate that equivalent fractions have the same magnitude. Seeing that two numbers share the same space on a number line may be unnerving for students (Izsák et al., 2008), but locating equivalent fractions on number lines or locating a single fraction when an equivalent fraction is already marked can help them to better understand the concept of equivalence, and that there are many fractions which occupy a given point on the number line (Siegler et al., 2010).

One final concern is that limitations in cognitive capacities such as verbal intelligence, attention, working memory, and executive functioning may interfere with student learning of mathematics generally, and fractions more specifically (Ye et al., 2016). The significance of each cognitive competency to mathematics is dynamic, shifting both by topical area (e.g., algebra, fraction knowledge, word problems) and by age (Bull et al., 2008; Jordan et al., 2013; Vukovic et al., 2014; Peng et al., 2019); even within fraction knowledge, these skills differ in predicting variance for factual knowledge, procedural skill, and conceptual understanding (Cragg et al., 2017). In particular, executive functioning measures—such as working memory (e.g., Baddeley & Hitch, 1974)—have been identified as explaining unique variance in math achievement and accuracy in middle school students (Siegler & Pyke, 2013), and the importance of working memory may become increasingly influential on more and more complex mathematics tasks (see Peng et al., 2019 for a review). Though working memory does contribute to mathematical achievement (Miller-Cotto & Byrnes, 2020), the relationship is partially mediated by domain-specific skills such as factual knowledge, procedural skill, and conceptual understanding (Cragg et al., 2017). Specifically for fractions, domain-specific mathematical skills may impact achievement more directly than domain-general skills (Hecht et al., 2003). Nevertheless, successful instruction for complex mathematics concepts and procedures, including those with fractions, should be designed to minimize barriers to learning that may be experienced by students with limitations in working memory or other cognitive capacities.

The Present Study

Findings from previous research about both domain-general and domain-specific factors that influence fraction knowledge yield several implications for how fractions should be taught. For instance, if students with limited cognitive capacities have difficulty learning about fractions (e.g., Cragg et al., 2017), improving their fraction understanding may require fraction instruction that is structured in a way that minimizes cognitive load. Generally, cognitive load is the amount of working memory resources one uses to solve a problem or complete a task. Further, if facets of both students' prior conceptual and procedural knowledge influence the degree to which they learn, successful fraction instruction should support the development of both conceptual and procedural knowledge.

In the present study, we aim to improve middle school students' fraction knowledge using an intervention strategy that simultaneously accomplishes all of these goals: Having students explain worked examples. Worked examples are completed math problems that display a fictitious student's solution to a problem, clearly marked correct or incorrect. Each example is accompanied by one or more prompts for students to notice and explain something about the features of the problem or the fictitious student's work (Booth et al., 2015a). Research has demonstrated that having students explain both correct and incorrect worked examples—that is, articulating concepts supporting correct examples or the misconceptions evident in incorrect examples—improves learning in mathematics classes (Barbieri & Booth, 2016), with impacts even greater for students who typically struggle with learning math (Booth et al., 2015b).

Worked examples are expected to be particularly useful for improving fraction learning as they inherently address two aforementioned difficulties that plague students learning about fractions. First, though students with limited cognitive capacities such as working memory may struggle with fractions and other complex mathematics, worked examples are widely known for reducing the extraneous (i.e., unnecessary) cognitive load that is found in mathematical problem-solving tasks (e.g., Sweller & Cooper, 1985). Further, students generally struggle with fractions if either their conceptual or procedural knowledge is insufficient. Compared to the more typical strategy of having students solve sets of practice problems on their own, interleaving worked examples and self-explanation within students' practice sessions is known to support both conceptual and procedural mathematics knowledge (e.g., Booth et al., 2015a).

The worked example approach targets common misconceptions about fraction problem features and incorporates visual representations—and fictitious students' interactions with these visual representations. Based on demonstrated success in targeting procedural skill and the conceptual underpinnings of those procedures (Booth et al., 2015a), we expect this to yield appreciable improvements in students' fraction computation skills. However, we also believe this approach can be effectively used to improve students' understanding of fractions as numbers alongside whole numbers on the number line, by addressing common misconceptions about the relative magnitudes of fractions. As this second hypothesis is more exploratory, we designed the present study to first examine the impact of the worked example approach on improving fraction computation knowledge, and then to determine if having students explain worked examples about fraction magnitude yielded additional benefit for student learning. Finally, we examine the relationship between learning each skill on the other. The literature would support that magnitude knowledge should affect the application of procedural knowledge in computation (Schneider et al., 2011) and we explore the relationship between computation knowledge and magnitude learning.

METHODS

We examined three research questions in this study. What is the impact of example-based worksheets for eighth graders' learning of fraction computation knowledge? What is the impact of example-based worksheets for eighth graders' learning of fraction magnitude knowledge? How do eighth graders' fraction magnitude and fraction computation knowledge relate to each other?

Participants

Participants were eighth grade students (N = 130; 61 males, 69 females) from 23 middle-school mathematics classrooms in four school districts in the Midwestern and Northeastern United States. The demographic breakdown of the participants was as follows: 69% White, 15% Black, 5% Asian, 5% Latinx, and 5% biracial or other. Individual data on the socio-economic status of the students was not available. The percentages of the populations of the four school districts who were eligible for free or reduced-cost lunch (i.e., a proxy for low-SES) were, 98% 58%, 30%, and 26%.

Individual students were randomly assigned to one of three conditions stratified by mathematics class: Control (Magnitude practice + Computation practice; n = 41), CompEx (Magnitude practice + Computational examples, n = 46), or Comp+MagEx (Magnitude examples + Computational examples, n = 43). A total of two students did not complete any posttest measures for the study and were therefore excluded from all analyses, leaving a study sample of 128 students. Individual students were further excluded from individual analyses if they did not complete the relevant outcome measure.

Measures

This study took place in the context of a larger project about connections between learning fractions and algebra; thus, all measures were administered along with other fraction or algebra measures which are not reported here. In order to examine the effectiveness of having students explain correct and incorrect worked examples of fraction computation and/or fraction magnitude problems, we used measures of both fraction computation and fraction magnitude knowledge.

Fraction Computation Knowledge. *Fraction computation knowledge* was measured using 12 fraction computation items like those that would typically be found in textbook lessons on fraction arithmetic. Students were assessed on fraction addition (2 items, e.g., 2/3 + 3/8), fraction subtraction (3 items, e.g., 6 2/5 – 2 4/5), fraction multiplication (4 items, e.g., 2 1/2 x 4), and fraction division (3 items, e.g., 2/9 ÷ 3/8). Each item was scored for correctness and the percent of items answered correctly was computed for each student. Internal consistency was good ($\alpha = .84$).

Fraction Magnitude Knowledge. A number line estimation task (e.g., Siegler & Opfer, 2003) was used to assess *fraction magnitude knowledge*. In this task, blank number lines with 0 marked at the left endpoint and 1 marked at the right endpoint were presented to students. Students were given a fraction with a magnitude between 0 and 1 and asked to place a mark on the number line where they believed that fraction would fall. Students received a total of eight number lines, each measuring 124 mm and staggered on a single page; the fractions they were asked to estimate were $\frac{3}{11}, \frac{1}{3}, \frac{7}{9}, \frac{2}{7}, \frac{1}{180}, \frac{1}{45}, \frac{1}{360}, \frac{5}{8}$. For each estimate, we measured how many millimeters were between the left (0) endpoint and the student's mark. We then determined which number actually belonged at the place where the student made their mark. Estimation accuracy was assessed using percent absolute error (PAE; Siegler & Booth, 2004), which was computed as (|Participant's estimate–Actual number|)/Numerical Range × 100%). Thus, if a student placed the fraction $\frac{1}{3}$ at the point where $\frac{1}{4}$ would actually go, their PAE for that item would be $|(.33 - .25)|/1 \times 100\%$, or 8%. We computed the mean PAE across all eight items for each student. Internal consistency was good ($\alpha = .85$). In addition, 30 number line tasks were double measured to check for systematic measurement errors; we found strong inter-rater reliability for PAE of number lines that were double measured ($K = .80$).

Intervention

Students in all conditions received 22 experimental tasks, each containing two items. In total, there were 12 worksheets (24 problems) targeting fraction magnitude knowledge and 10 worksheets (20 problems) targeting fraction computation knowledge. Within the fraction computation worksheets, there were four different topic areas: Fraction Addition (2 worksheets; 4 problems), Fraction Subtraction (2 worksheets; 4 problems), Fraction Multiplication (2 worksheets; 4 problems), and Fraction Division

(4 worksheets; 8 problems). The three conditions varied only in whether their fraction computation and/or fraction magnitude worksheets consisted only of practice problems, or whether they included worked examples, too. For practice-based worksheets, students simply solved two practice problems on their own. For example-based worksheets, one of the two practice problems was replaced with an example of a fictitious student's correct or incorrect work to solve the problem—clearly marked as correct or incorrect—and prompts for students to explain a problem feature or concept in the example. Then, they solved the paired practice problem on their own. Because we used within-classroom random assignment, the teachers' instruction and all other student assignments during the class did not vary by condition.

Control Condition: Practice Only. In the Practice-Only condition, students received practice-based worksheets for both fraction magnitude and fraction computation. Thus, they completed 24 magnitude estimation practice problems, and 20 computation practice problems to solve on their own (see sample worksheets in Figure 7.1).

Experimental Condition 1: Fraction Computation Worked Examples. In the Computation Examples (CompEx) condition, students received

Figure 7.1

Sample Magnitude and Computation Worksheets for Control Condition

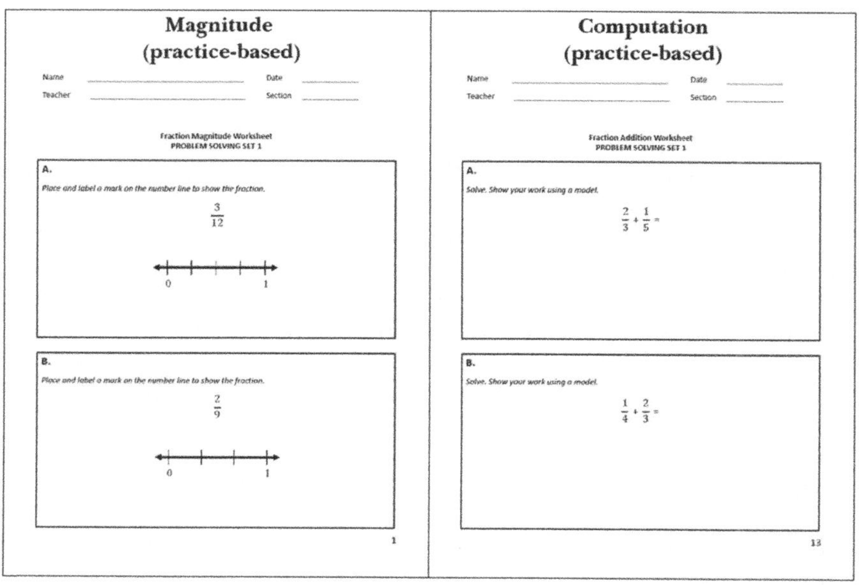

practice-based worksheets for fraction magnitudes and example-based worksheets for fraction computation. Thus, they completed 24 magnitude estimation practice problems, and 10 fraction arithmetic worked examples (5 correct, 5 incorrect) with self-explanation prompts, and 10 fraction computation practice problems (see sample worksheets in Figure 7.2).

Figure 7.2

Sample Magnitude and Computation Worksheets for CompEx Condition

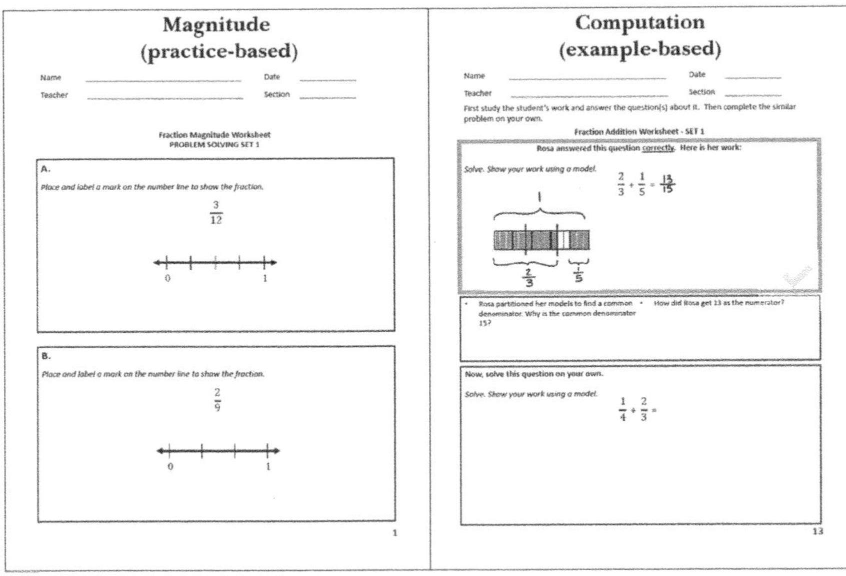

Experimental Condition 2: Fraction Computation and Fraction Magnitude Examples. In the Computation and Magnitude Examples (Comp+MagEx) condition, students received example-based worksheets for both fraction magnitude and fraction computation. Thus, they completed 12 fraction magnitude worked examples (6 correct and 6 incorrect) with self-explanation prompts, 12 fraction magnitude practice problems, 10 fraction arithmetic worked examples (5 correct, 5 incorrect) with self-explanation prompts, and 10 fraction computation practice problems (see sample worksheets in Figure 7.3).

Procedure

Individual students in participating classrooms were randomly assigned to one of the three conditions (Control, CompEx, or Comp+MagEx).

Figure 7.3

Sample Magnitude and Computation Worksheets for Comp+MagEx Condition

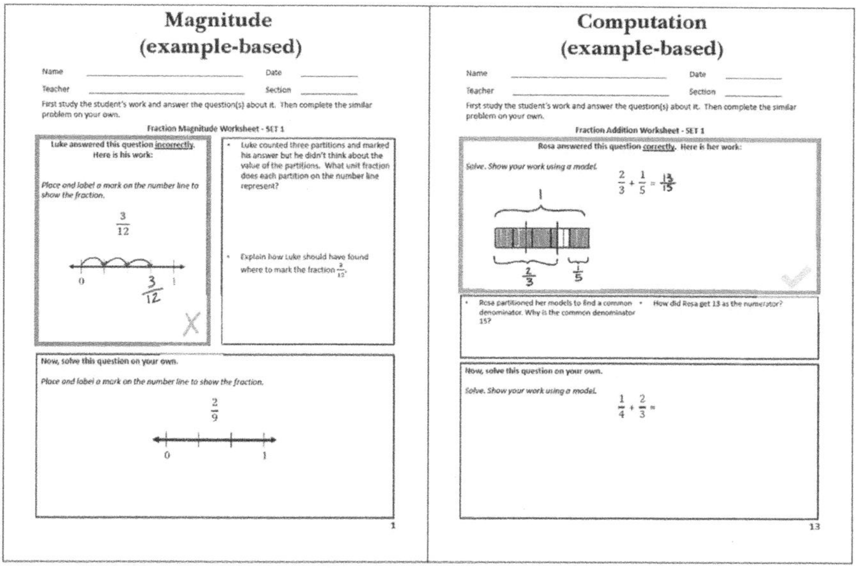

Teachers administered the study measures as a pretest during the first marking period, before beginning any study worksheets. Then, students were given packets of worksheets that corresponded to their assigned condition. During the typical, teacher-provided instruction or review on fractions, teachers were asked to use all of the worksheets and to assign individual study worksheets at times that made sense in their classroom. Following completion of all study worksheets, teachers administered the measures again as an immediate posttest to their students; the timing of this immediate posttest varied by classroom depending on how the teacher spread out the assignments. Then, near the end of the final marking period, teachers again administered the measures as a delayed posttest, and the school provided demographic data (gender, race, and ethnicity) of individual students.

RESULTS

In the following section we present our data analysis plan. The subsequent sections detail the findings for each research question. Finally, we discuss how these results might impact future educational considerations and potential research.

Data Analysis Plan

All three research questions were addressed at two time points: Immediate posttest and delayed posttest. Due to the nested nature of the data (students in classrooms), intra-class correlations were computed for immediate and delayed posttest PAE and computation measures; depending on the particular measure and time point, between 0% and 27% of the variance was accounted for at the classroom level. Multilevel modeling (MLM) was therefore used for all analyses to account for clustering. To assess the impact of example-based assignments on fraction computation skills at immediate and delayed posttest (i.e., Research Question 1), we conducted two MLMs (one for immediate posttest and one for delayed posttest) of posttest fraction computation scores with CompEx and Comp+MagEx conditions as dummy-coded predictors, controlling for pretest fraction computation scores. We used the same approach to assess the impact on fraction magnitude understanding (i.e., Research Question 2), predicting posttest PAE with CompEx and Comp+MagEx conditions as dummy-coded predictors, controlling for pretest PAE. Then, to assess the relationship between fraction magnitude and computation at immediate and delayed posttest (i.e., Research Question 3), we computed two parallel models. To determine whether improvements in PAE were related to improvements in computation scores, we first predicted immediate posttest computation using CompEx, Comp+MagEx, and immediate posttest PAE, controlling for pretest computation scores and pretest PAE, and then we predicted delayed posttest computation using CompEx, Comp+MagEx, and delayed posttest PAE, controlling for pretest computation scores and pretest PAE. To determine whether improvements in fraction computation scores impacted improvements in PAE, we predicted immediate posttest PAE using CompEx, Comp+MagEx, and immediate posttest computation scores, controlling for pretest computation scores and pretest PAE, and then predicted delayed posttest PAE using CompEx, Comp+MagEx, and delayed posttest computation scores, controlling for pretest computation scores and pretest PAE. Means by condition for all pretest and posttest measures are displayed in Table 7.1.

Research Question 1: Impact of Example-Based Worksheets on Fraction Computation

As shown in Table 7.2, fraction computation was not impacted by the experimental conditions at immediate posttest (Model 1) or delayed posttest (Model 2). That is, neither students in the CompEx or Comp+MagEx conditions (both of which completed worksheets with worked examples for

Table 7.1

Mean Pretest and Posttest Scores by Condition

	Fraction Magnitude (PAE)			Fraction Computation		
Condition	**Pretest**	**Posttest**	**Delayed Posttest**	**Pretest**	**Posttest**	**Delayed Posttest**
Control	0.18	0.15	0.18	0.43	0.51	0.39
CompEx	0.17	0.13	0.12	0.51	0.58	0.47
Comp+MagEx	0.10	0.06	0.06	0.51	0.57	0.53

Table 7.2

Effect f Condition on Fraction Computation at Immediate and Delayed Posttest

Effect	**Parameter**	**Model 1: Immediate Posttest**	**Model 2: Delayed Posttest**
		Fixed Effects	Fixed Effects
(Intercept)	Y_{00}	0.23 (.06)	0.11 (.06)
CompEx	Y_{10}	-0.01 (.05)	0.01 (.05)
Comp+MagEx	Y_{11}	-0.02 (.05)	0.04 (.05)
Pretest Computation	Y_{12}	0.67 (.08)***	0.69 (.08)***
		Random Effects	Random Effects
Variance component			
Class		0.01	0.01705
Residual		0.45	0.04531
		Goodness of Fit	Goodness of Fit
AIC		11.49	15.7191
BIC		28.06	32.1887
R^2 (fixed effects)		0.41	0.4166
R^2 (total)		0.54	0.5761

***$p < .001$, **$p < .01$, *$p < .05$

fraction computation) improved their fraction computation any more than students in the control condition.

Research Question 2: Impact of Example-Based Worksheets on Fraction Magnitude

As shown in Table 7.3, immediate posttest PAE was not differentially impacted by the experimental conditions (Model 1) in comparison to the control condition, but there was a significant effect of the Comp+MagEx condition (worked examples for computation + magnitude) for delayed posttest PAE. Students in the Comp+MagEx condition improved their fraction magnitude understanding (i.e., reduced their PAE) by the end of the year more than students in the Control condition. There was not a significant effect of the CompEx condition (worked examples for computation only) on PAE compared to the control condition.

Table 7.3

Effect of Condition on Fraction Magnitude Learning at Immediate and Delayed Posttest

Effect	Parameter	Model 1: Immediate Posttest	Model 2: Delayed Posttest
		Fixed Effects	Fixed Effects
(Intercept)	Y_{00}	0.06 (.02)	0.11 (.03)
CompEx	Y_{10}	–0.01 (.03)	–0.05 (.03)
Comp+MagEx	Y_{11}	–0.03 (.03)	–0.08 (.04)**
Pretest PAE	Y_{12}	0.45 (.07)***	0.36(.08)***
		Random Effects	Random Effects
Variance component			
Class		0.00	0
Residual		0.01	0.1268
		Goodness of Fit	Goodness of Fit
AIC		–127.73	–89.3792
BIC		–112.28	–74.314
R^2 (fixed effects)		0.30	0.2262
R^2 (total)		0.30	0.2262

***$p < .001$, **$p < .01$, *$p < .05$

Research Question 3: Effects of Magnitude on Computation and Computation on Magnitude

As shown in Table 7.4, students' immediate posttest PAE was negatively related to their immediate posttest computation knowledge, controlling for their pretest PAE. This indicates that the better students' posttest fraction magnitude performance (less error) by the immediate posttest, the higher students' computation knowledge at the same time point. Delayed posttest PAE was not related to delayed posttest computation knowledge. This makes sense because, though PAE improvements between pre- and immediate posttest were either retained or continued to improve by delayed posttest overall, fraction computation skills showed declines between immediate and delayed posttest.

Table 7.4

Effect of Fraction Magnitude Knowledge on Computation at Immediate and Delayed Posttest

Model 1: PAE at Immediate Posttest			**Model 2: PAE at Delayed Posttest**		
Effect	**Parameter**		**Effect**	**Parameter**	
		Fixed Effects			Fixed Effects
(Intercept)	Y_{00}	0.26 (.06)	(Intercept)	Y_{00}	0.17 (.09)
CompEx	Y_{10}	0.00 (.05)	CompEx	Y_{10}	0.00 (.06)
Comp+MagEx	Y_{11}	–0.04 (.05)	Comp+MagEx	Y_{11}	0.02 (.07)
Pretest Computation	Y_{12}	0.72 (.08)***	Pretest Computation	Y_{12}	0.68 (.10)***
Pretest PAE	Y_{13}	0.19 (.17)	Pretest PAE	Y_{13}	0.05(.18)
Immediate Posttest PAE	Y_{14}	–0.61 (.20)**	Delayed Posttest PAE	Y_{15}	-0.36 (.19)
		Random Effects			Random Effects
Variance component					
ClassID		0.009751			0.0154
Residual		0.039606			0.04807
		Goodness of Fit			Goodness of Fit
AIC		5.8313			26.7536
BIC		26.429			46.7521
R^2 (fixed effects)		0.5038			0.2505
R^2 (total)		0.6018			0.5641

$^{***}p < .001$, $^{**}p < .01$, $^{*}p < .05$

As shown in Table 7.5, after controlling for their pretest fraction computation knowledge, students' immediate posttest computation knowledge was negatively related to their immediate posttest PAE. This indicates that the higher students' computation scores at immediate posttest, the lower percent absolute error (or higher accuracy) on the number line task. However, delayed posttest computation knowledge was not related to delayed posttest PAE, consistent with the prior model.

Table 7.5

Effect of Fraction Computation Knowledge on PAE at Immediate and Delayed Posttest

Model 1: Computation at Immediate Posttest			**Model 2: Computation at Delayed Posttest**		
Effect	**Parameter**		**Effect**	**Parameter**	
		Fixed Effects			Fixed Effects
(Intercept)	Y_{00}	0.08 (.03)	**(Intercept)**	Y_{00}	0.13 (.04)
CompEx	Y_{10}	–0.01 (.03)	**CompEx**	Y_{10}	–0.05 (.03)
Comp+MagEx	Y_{11}	–0.03 (.03)	**Comp+MagEx**	Y_{11}	–0.08 (.04)*
Pretest PAE	Y_{12}	0.44 (.07)***	**Pretest PAE**	Y_{12}	0.34(.09)***
Pretest Computation	Y_{13}	0.11 (.05)*	**Pretest Computation**	Y_{13}	0.06 (.06)
Immediate Posttest Computation	Y_{14}	–0.14 (.05)**	**Delayed Posttest computation**	Y_{15}	-0.10 (.05)
		Random Effects			Random Effects
Variance component					
ClassID		0			0
Residual		0.0108			0.1597
		Goodness of Fit			Goodness of Fit
AIC		–123.0228			–78.4781
BIC		–102.4251			–58.4797
R^2 (fixed effects)		0.3532			0.2505
R^2 (total)		0.3532			0.2505

$^{***}p < .001$, $^{**}p < .01$, $^{*}p < .05$

DISCUSSION

In the present study, we examined the effectiveness of having eighth grade students explain correct and incorrect worked examples to improve their fraction computation and fraction magnitude knowledge; we also examined the reciprocal impacts of fraction computation knowledge and fraction magnitude knowledge on each other's development. Findings from the present study indicated that the worked example approach was more effective than a problem-solving control at improving fraction magnitude knowledge but no more or less effective than the control at improving fraction computation skills.

Influences on Fraction Magnitude Knowledge

To our knowledge, this study was the first to examine whether explaining worked examples of fictitious students' completion of fraction magnitude problems yields improvements in students' fraction number line estimation acuity. Given recommendations that students must recognize that fractions are numbers and that they expand the number system beyond whole numbers (Siegler et al., 2010), finding ways to improve students' magnitude knowledge in this way is important. In the present study, students in the Control group and the CompEx group practiced solving fraction magnitude problems on their own, while students in the Comp+MagEx condition were prompted to explain worked examples in place of half of the problems to solve. Results indicate that the fraction magnitude examples in Comp+MagEx condition were indeed effective for improving students' fraction magnitude knowledge. Interestingly, the difference between conditions was not significant on the immediate posttest, right after the intervention was complete, but emerged over time as the Comp+MagEx group continued to improve, and the difference was evident on the delayed posttest at the end of the school year. However, the fraction computation worked examples received by students in the CompEx condition did not yield any significant differences in fraction magnitude scores compared to the problem-solving control. That is, worked examples designed to help students better understand the concepts underlying computational procedures with fractions did not produce improvements in their understanding of fraction magnitudes.

Fraction magnitude knowledge was related to students' fraction computation knowledge in two ways. First, students with higher fraction computation scores at pretest had lower fraction magnitude scores at immediate posttest, controlling for pretest PAE. Second, students with higher fraction computation scores at immediate posttest tended to have higher fraction magnitude knowledge at immediate posttest. The second finding supports prior findings from observational work with fourth through sixth graders that fraction magnitude knowledge and fraction computations develop iteratively (Bailey et al., 2017). This echoes findings that conceptual knowledge and procedural knowledge interact in an iterative fashion (Rittle-Johnson & Alibali, 1999). The finding that students with higher pretest fraction computation knowledge tend to have lower fraction magnitude knowledge at immediate posttest is perhaps more interesting, as it indicates that having a stronger foundation of fraction computation knowledge at the beginning of the year may not afford students with benefits for learning about fraction magnitudes across the year. This contrast could be because, by eighth grade, students have already experienced a considerable amount of fraction procedural instruction that occurred without regard for the magnitudes of those fractions; thus that knowledge is not helpful in this way. However, since the fraction computation instruction and/or practice that students received in the present study was presented in conjunction with instruction and practice on fraction magnitude work, the newly gained fraction computation knowledge may support fraction magnitude learning. Further work is needed to tease apart how and when performance and learning of fraction computation might have an impact on the development of students' fraction magnitude knowledge.

Influences on Fraction Computation Knowledge

Contrary to decades of research in laboratories and more recent studies in classrooms showing the effectiveness of worked examples for students learning to solve problems in mathematics (see Booth et al., 2017), the worked example intervention in the present study yielded no significant improvements in students' fraction computation skills compared to a problem-solving control. Students in both the CompEx and Comp+MagEx conditions explained correct and incorrect worked examples about fraction computation, particularly focusing their attention on the concepts that underlie the use of the fraction computation procedures being demonstrated. However, students in those conditions did not outperform Control students who simply solved the fraction computation problems on their own; this was the case at both the immediate and delayed posttests. One possible reason for this discrepant finding could be that the eighth grade students in the study had learned this content in previous grades and the teachers were using these materials for a review. Some evidence suggests that the worked example approach is especially effective when students are just beginning to learn content (Kalyuga et al., 2001). However, there is also evidence that worked examples are particularly effective for students who struggle with the relevant content (Barbieri & Booth, 2016, 2020). Given that fraction computation has proven to be a topic with which even adults can struggle (Newton, 2008), one would expect that the worked examples could still be beneficial in middle school. Further work would be needed to examine how individual differences in prior knowledge might impact the effectiveness of the approach. It is also possible that the worked example approach is just less effective for fraction computation compared with other mathematical content; future research should also investigate differences in the effectiveness of this approach for different types of mathematical content.

Though the worked example intervention did not differentially impact fraction computation, it is interesting to note that an important relation was found between students' fraction magnitude knowledge and fraction computation at immediate posttest: Even after controlling for students' pretest magnitude knowledge, students who had lower PAE on their fraction number lines after the intervention was complete tended to also have higher fraction computation scores at posttest. This suggests the importance of fraction magnitude knowledge in performance and learning for other mathematics skills and highlights the interconnected development of conceptual and procedural knowledge in mathematics (Rittle-Johnson et al., 2001).

Educational Implications

One key implication of the present study is that while the worked example approach did not help or hurt students' computational skill with fractions in eighth grade, it did positively impact their knowledge of fraction magnitude as measured by a delayed posttest. Magnitude knowledge is critical for mathematics learning, and fraction magnitude knowledge in particular supports readiness for algebra (Booth & Newton, 2012). Results from this study underscore the need for remediation of fraction magnitude knowledge during the early learning of algebra. This pivotal point of learning fractions could be where some students take an exit ramp from mathematics rather than the path that "leads students gently from arithmetic up to algebra" described by (Wu, 2001, p. 1). Key to this early conceptual understanding of magnitude is a scalable intervention which can be implemented nationally. The worked example worksheets have proven to be a successful tool that meets the criteria for overburdened teachers and school systems without a great deal of financial resources. However, when it comes to fractions, it may be a more effective solution for students in elementary school or earlier in middle school but may not be as effective for remediation of fraction computational knowledge for older students who have already learned the content via more typical mathematics instruction. The understanding of the magnitude of fractions, as numbers which exist between whole numbers, is their only option to "scale the wall of algebra" (Wu, 2001, p. 1). This research suggests that our focus for remediation should be on fraction magnitude.

One of the limitations of this study is that due to within-class random assignment, teachers could not review or provide feedback on the worked example problems with their students; recall that while students in the Comp+MagEx condition completed examples for both magnitude and computation items, students in the CompEx condition only had examples for computation items and control students did not have any worked examples. Previous work by Fazio and colleagues (2016) exploring the effects of conceptual magnitude knowledge on improving fraction comparison emphasized that meaningful feedback on errors was key to improvement in fraction comparison. In the present study, worked examples provided students with some feedback in the form of drawing attention to errors made by fictitious students; however, no students received individualized feedback on their responses. Future research using worked examples as a remediation practice for students who struggle with fraction magnitude might examine teacher feedback as a possible additional factor in the success of this intervention.

Finally, while this study suggests that it may still be beneficial to remediate fraction magnitude knowledge as late as middle school, it is yet unclear whether improvements in fraction magnitude knowledge at this point are still able to provide a boost for students' mathematics learning, or at which point such improvements yield appreciable transfer effects, for example, to algebra. Further research is needed to determine if there is a point at which students must have already gained a solid foundation of fraction magnitude knowledge before moving onto higher mathematics; studies on developmental effects of the impact of this type of treatment would help us to know whether, when, and how intervening on fraction magnitude knowledge is effective.

CONCLUSION

Findings from the present study indicated that the worked example approach was more effective than a problem-solving control at improving fraction magnitude knowledge but no more or less effective than the control at improving fraction computation skills. Results indicate that the fraction magnitude examples in Comp+MagEx condition were indeed effective for improving students' fraction magnitude knowledge. However, the fraction computation worked examples received by students in the CompEx condition did not yield any significant differences in fraction magnitude scores compared to the problem-solving control. That is, worked examples designed to help students better understand the concepts underlying computational procedures with fractions did not produce improvements in their understanding of fraction magnitudes.

ACKNOWLEDGMENTS

Acknowledgements: Funding for this research was provided by the Institute of Education Sciences, U.S. Department of Education, through Grant R305A170226 to Temple University. The opinions expressed are those of the authors and do not represent the views of the Institute or U.S. Department of Education.

REFERENCES

Anand, P. G., & Ross, S. M. (1987). Using computer-assisted instruction to personalize arithmetic materials for elementary school children. *Journal of Educational Psychology*, *79*(1), 72–78. https://doi.org/ 10.1037/0022-0663.79.1.72

Baddeley, A. D., & Hitch, G. (1974). Working memory. *Psychology of Learning and Motivation*, *8*, 47–89. https://doi.org/10.1016/S0079-7421(08)60452-1

Bailey, D. H., Hansen, N., & Jordan, N. C. (2017). The codevelopment of children's fraction arithmetic skill and fraction magnitude understanding. *Journal of Educational Psychology*, *109*(4), 509–519. https://doi.org/10.1037/edu0000152

Barbieri, C., & Booth, J. L. (2016). Support for struggling students in algebra: Contributions of incorrect worked examples. *Learning and Individual Differences*, *48*, 36–44. https://doi.org/10.1016/j.lindif.2016.04.001

Barbieri, C. A., & Booth, J. L. (2020). Mistakes on display: Incorrect examples refine equation solving and algebraic feature knowledge. *Applied Cognitive Psychology*, *34*(4), 862–878. https://doi.org/10.1002/acp.3663

Barbieri, C. A., Rodrigues, J., Dyson, N., & Jordan, N. C. (2020). Improving fraction understanding in sixth graders with mathematics difficulties: Effects of a number line approach combined with cognitive learning strategies. *Journal of Educational Psychology*, *112*(3), 628–648. https://doi.org/10.1037/edu0000384

Barbieri, C., Young, L., Newton, K. J., & Booth, J. L. (in press). Predicting middle school profiles of algebra performance using fraction knowledge. *Child Development*.

Beckmann, S. (2008). *Mathematics for elementary teachers and activities* (2nd ed.). Addison Wesley.

Booth, J. L., & Koedinger, K. R. (2008). Key misconceptions in algebraic problem solving. *Proceedings of the Annual Meeting of the Cognitive Science Society*, *30*, 571–576. https://escholarship.org/uc/item/5n28t12n

Booth, J. L., McGinn, K. M., Barbieri, C., Begolli, K., Chang, B, Miller-Cotto, D., Young, L. K., & Davenport, J.L. (2017). Evidence for cognitive science principles that impact learning in mathematics. In D. C. Geary, D. Berch, R. Oschendorf, & K. M. Koepke (Eds.), *Mathematical cognition and learning: Acquisition of complex arithmetic skills and higher-order mathematics concepts* (Vol. 3, pp. 297–325). Elsevier.

Booth, J. L., & Newton, K. J. (2012). Fractions: Could they really be the gatekeeper's doorman? *Contemporary Educational Psychology, 37*(4), 247–253. https://doi.org/10.1016/j.cedpsych.2012.07.001

Booth, J. L., Cooper, L. A., Donovan, M. S., Huyghe, A., Koedinger, K. R., & Paré-Blagoev, E. J. (2015a). Design-Based Research Within the Constraints of Practice: AlgebraByExample. *Journal of Education for Students Placed at Risk, 20*(1-2), 79–100. https://doi.org/10.1080/10824669.2014.986674

Booth, J. L., Oyer, M. H., Paré-Blagoev, E. J., Elliot, A. J., Barbieri, C., Augustine, A., & Koedinger, K. R. (2015b). Learning algebra by example in real-world classrooms. *Journal of Research on Educational Effectiveness, 8*(4), 530–551. https://doi.org/10.1080/19345747.2015.1055636

Brown, G., & Quinn, R. J. (2006). Algebra students' difficulty with fractions: An error analysis Australian mathematics teacher. *Australian Mathematics Teacher, 62*(4), 28–40.

Bull, R., Espy, K. A., & Wiebe, S. A. (2008). Short-term memory, working memory, and executive functioning in preschoolers: Longitudinal predictors of mathematical achievement at age 7 years. *Developmental Neuropsychology, 33*(3), 205–228. https://doi.org/10.1080/87565640801982312

Cragg, L., Keeble, S., Richardson, S., Roome, H. E., & Gilmore, C. (2017). Direct and indirect influences of executive functions on mathematics achievement. *Cognition, 162*, 12–26. https://doi.org/10.1016/j.cognition.2017.01.014

Cramer, K. A., Post, T. R., & Delmas, R. C. (2002). Initial fraction learning by fourth- and fifth-grade students: A comparison of the effects of using commercial curricula with the effects of using the rational number project curriculum. *Journal for Research in Mathematics Education, 33*(2), 111–144. https://doi.org/10.2307/749646

Crooks, N. M., & Alibali, M. W. (2014). Defining and measuring conceptual knowledge in mathematics. *Developmental Review, 34*(4), 344–377. https://doi.org/10.1016/j.dr.2014.10.001

Fazio, L. K., Kennedy, C. A., & Siegler, R. S. (2016). Improving children's knowledge of fraction magnitudes. *PLoS ONE, 11*(10), 1–15. https://doi.org/10.1371/journal.pone.0165243

Hecht, S. A., Close, L., & Santisi, M. (2003). Sources of individual differences in fraction skills. *Journal of Experimental Child Psychology, 84*(3), 277–302. https://doi.org/10.1016/j.jecp.2003.08.003

Hudson Hawkins, V. (2008). *The effects of math manipulatives on student achievement in mathematics* [Unpublished doctoral dissertation, Cappella University].

Institute of Education Sciences. (2005) *Funding opportunities | Search funded research grants and contracts*. https://ies.ed.gov/funding/grantsearch/

Irwin, K. C. (2001). Using everyday knowledge of decimals to enhance understanding. *Journal for Research in Mathematics Education, 32*(4), 399–420. https://doi.org/10.2307/749701

Izsák, A., Tillema, E., & Tunç-Pekkan, Z. (2008). Teaching and learning fraction addition on number lines. *Journal for Research in Mathematics Education, 39*(1), 33–62.

Jordan, N. C., Hansen, N., Fuchs, L. S., Siegler, R. S., Gersten, R., & Micklos, D. (2013). Developmental predictors of fraction concepts and procedures. *Journal of Experimental Child Psychology, 116*(1), 45–58. https://doi.org/10.1016/j.jecp.2013.02.001

Kalyuga, S., Chandler, P., Tuovinen, J., & Sweller, J. (2001). When problem solving is superior to studying worked examples. *Journal of Educational Psychology, 93*, 579–588.

Keijzer, R., & Terwel, J. (2003). Learning for mathematical insight: A longitudinal comparative study on modelling. *Learning and Instruction, 13*(3), 285–304. https://doi.org/10.1016/S0959-4752(02)00003-8

Kerslake, D. (1986). *Fractions: Children's strategies and errors. A report of the strategies and errors in secondary mathematics project.* NFER-Nelson.

Lee, S. W., & Mao, X. (2020). Algebra by the Eighth Grade: The association between early study of algebra I and students' academic success. *International Journal of Science and Mathematics Education*. https://doi.org/10.1007/s10763-020-10116-3

Mack, N. (1995). Confounding whole-number and fraction concepts when building on informal knowledge. *Journal for Research in Mathematics Education, 26*, 422–441. https://doi.org/10.2307/749431

Mack, N. K. (2000). Long-term effects of building on informal knowledge in a complex content domain: The case of multiplication of fractions. *Journal of Mathematical Behavior, 19*(3), 307–332. https://doi.org/10.1016/s0732-3123(00)00050-x

Miller-Cotto, D., & Byrnes, J. P. (2020). What's the best way to characterize the relationship between working memory and achievement?: An initial examination of competing theories. *Journal of Educational Psychology, 112*(5), 1074–1084. https://doi.org/10.1037/edu0000395

Moseley, B. J., Okamoto, Y., & Ishida, J. (2007). Comparing US and Japanese elementary school teachers' facility for linking rational number representations. *International Journal of Science and Mathematics Education, 5*(1), 165–185. https://doi.org/10.1007/s10763-006-9040-0

National Science Foundation. (n.d.). *Awards simple search*. https://www.nsf.gov/awardsearch/

Newton, K. J. (2008). An extensive analysis of preservice elementary teachers' knowledge of fractions. *American Educational Research Journal, 45*(4), 1080–1110.

Nishida, T. K. (2008). *The use of manipulatives to support children's acquisition of abstract math concepts* [Unpublished dissertation, University of Virginia].

Painter, R. R. (1989). *A comparison of the procedural error patterns, scores, and other variables, of selected groups of university and eighth grade students in Mississippi on a test involving arithmetic operations on fractions* [Unpublished thesis, University of Southern Mississippi].

Peng, P., Wang, T., Wang, C., & Lin, X. (2019). A meta-analysis on the relation between fluid intelligence and reading/mathematics: Effects of tasks, age, and social economics status. *Psychological Bulletin, 145*(2), 189–236. https://doi.org/10.1037/bul0000182

Rittle-Johnson, B., & Alibali, M. W. (1999). Conceptual and procedural knowledge of mathematics: Does one lead to the other? *Journal of Educational Psychology, 91*(1), 175–189. https://doi.org/10.1037/0022-0663.91.1.175

Rittle-Johnson, B., & Schneider, M. (2015). *Developing conceptual and procedural knowledge of mathematics*. In R. C. Kadosh & A. Dowker (Eds.), *Oxford library of psychology. The Oxford handbook of numerical cognition* (pp. 1118–1134). Oxford University Press. https://doi.org/10.1093/oxfordhb/9780199642342.013.014

Rittle-Johnson, B., & Siegler, R. S. (1998). The relation between conceptual and procedural knowledge in learning mathematics: A review. In C. Donlan (Ed.), *Studies in developmental psychology. The development of mathematical skills* (pp. 75–110). Psychology Press/Taylor & Francis.

Rittle-Johnson, B., Siegler, R. S., & Alibali, M. W. (2001). Developing conceptual understanding and procedural skill in mathematics: An iterative process. *Journal of Educational Psychology, 93*(2), 346–362. https://doi.org/10.1037/0022-0663.93.2.346

Rodrigues, J., Dyson, N. I., Hansen, N., & Jordan, N. C. (2016). Preparing for Algebra by Building Fraction Sense. *TEACHING Exceptional Children, 49*(2), 134–141. https://doi.org/10.1177/0040059916674326

Schneider, M., Grabner, R. H., & Paetsch, J. (2009). Mental number line, number line estimation, and mathematical achievement: Their interrelations in grades 5 and 6. *Journal of Educational Psychology, 101*(2), 359–372. https://doi.org/10.1037/a0013840

Schneider, M., Rittle-Johnson, B., & Star, J. R. (2011). Relations among conceptual knowledge, procedural knowledge, and procedural flexibility in two samples differing in prior knowledge. *Developmental Psychology, 47*(6), 1525–1538. https://doi.org/10.1037/a0024997

Siegler, R., Carpenter, T., Fennell, F. S., Geary, D., Lewis, J., Okamoto, Y., Thompson, L., & Wray, J. (2010). *Developing effective fractions instruction for kindergarten through 8th grade: A practice guide*. National Center for Education Evaluation and Regional Assistance, Institute of Education Sciences, U.S. Department of Education. https://ies.ed.gov/ncee/wwc/Docs/PracticeGuide/fractions_pg_093010.pdf

Siegler, R. S., & Booth, J. L. (2004). Development of numerical estimation in young children. *Child Development, 75*(2), 428–444. https://doi.org/10.1111/j.1467-8624.2004.00684.x

Siegler, R. S., & Opfer, J. (2003). The development of numerical estimation: Evidence for multiple representations of numerical quantity. *Psychological Science, 14*, 237–243. https://doi.org/10.1111/1467- 9280.02438

Siegler, R. S., & Pyke, A. A. (2013). Developmental and individual difference in understanding of fractions. *Developmental Psychology, 49*(10), 1–22. https://doi.org/10.1038/jid.2014.371

Siegler, R. S., Duncan, G. J., Davis-Kean, P. E., Duckworth, K., Claessens, A., Engel, M., Susperreguy, M. I., & Chen, M. (2012). Early predictors of high school mathematics achievement. *Psychological Science, 23*(7), 691–697. https://doi.org/10.1177/0956797612440101

Sowder, J. T., Philipp, R. A., Armstrong, B. E., & Schappelle, B. P. (1998). *Middle grade teachers' mathematical knowledge and its relationship to instruction*. State University of New York Press.

Sweller, J., & Cooper, G. A. (1985). The use of worked examples as a substitute for problem solving in learning algebra. *Cognition and Instruction, 2*(1), 59–89. https://doi.org/10.1207/s1532690xci0201_3

U.S. Department of Education. (2008). *The Final Report of the National Mathematics Advisory Panel*. U.S. Department of Education. https://www2.ed.gov/about/bdscomm/list/mathpanel/report/final-report.pdf

Van de Walle, J., Karp, K., & Bay-Williams, J. (2010). *Elementary and middle school mathematics: Teaching developmentally* (7th ed.). Allyn and Bacon.

Vukovic, R. K., Fuchs, L. S., Geary, D. C., Jordan, N. C., Gersten, R., & Siegler, R. S. (2014). Sources of individual differences in children's understanding of fractions. *Child Development, 85*(4), 1461–1476. https://doi.org/10.1111/cdev.12218

Wu, H. (2001). How to prepare students for algebra. *American Educator, 25*(2), 10–17.

Ye, A., Resnick, I., Hansen, N., Rodrigues, J., Rinne, L., & Jordan, N. C. (2016). Pathways to fraction learning: Numerical abilities mediate the relation between early cognitive competencies and later fraction knowledge. *Journal of Experimental Child Psychology, 152*, 242–263. https://doi.org/10.1016/j.jecp.2016.08.001

CHAPTER 8

EXPANDING A COLLECTIVE ZONE OF PROXIMAL DEVELOPMENT

Investigating the Potential of Zankov's Lesson With Elementary Students Analyzing Numerical Equalities

Åsmund Gjære and Viktor Freiman

INTRODUCTION

Starting up a lesson on some algebraic ideas with a group of Grade 6 students from New Brunswick, Canada, the authors decided to begin with a small warm-up task with three numerical equalities, asking the students whether they are *true* or *false*:

$$24 \times 214 = 12 \times 428 \qquad \text{(I)}$$
$$325 + 427 = 529 + 225 \qquad \text{(II)}$$
$$6 \times 99 = 600 - 5 \qquad \text{(III)}$$

Knowing that the development of students' awareness about the relational nature of the equal sign is one of the key elements of the algebra strand

Mathematics as the Science of Patterns:
Making the Invisible Visible to Students Through Teaching, pp. 153–184

starting from early grades, we thought that students would see the first equality as *true* since 24 is double of 12 and 214 is half of 428.

We also expected students to look into the differences between 325 and 225 and between 427 and 529 and see that the expression on the right side is bigger than the one on the left side by 2, so the second equality is *false*. Finally, regarding the third equality, we could expect that some students would use the distributive property: $6 \times 99 = 6 \times 100 - 6 \times 1$ to find that the left side is smaller by 1, so equality III is also *false*.

To our surprise, all students enthusiastically began calculating the numerical values of the expressions on both sides of each equality. They used a variety of strategies, some quite sophisticated, and many succeeded to get correct answers. Nevertheless, our original intention to use these tasks as prelude to more complex algebraic reasoning was not fulfilled. As a result, an unplanned teaching-learning situation arose on the spot prompting some very rich follow-up discussions engaging students in a deeper investigation.

For some students, we could observe the emergence of different perspectives of "seeing" the mathematical structures of the equalities. This led the students toward previously unexpected learning paths, allowing us to investigate their developmental potential for further growth in algebraic thinking. The authors, in this chapter, discuss this classroom case from the theoretical perspective of the Vygotskian scholar L. V. Zankov (1901–1977).

Students' Difficulty With the Algebraic Perspective

A rich body of early algebra research developed in 1990–2000s has shown that students have difficulties perceiving the relational nature of numerical equalities, instead seeing them as chains of operations (e.g., Molina & Ambrose, 2008). For instance, one of the latest reports based on the strategies used by Korean children from Grade 3 showed that while students were able to use strategies based on the analysis of (algebraic) structure, they still preferred computations unless explicitly asked to evaluate equalities without calculations (Pang & Kim, 2018). Therefore, the search for efficient teaching approaches fostering young learners' capacity to develop a more relational perspective on arithmetic remains a current problem for research and practice (Kieran, 2018).

At the same time, in Western scholarship, there is a growing interest in the experimental curricula developed in the 1950–1960s by Soviet psychologists and grounded in a Vygotskian developmental line of thought (Freiman & Fellus, 2021). For instance, in Canada, a longitudinal study conducted by Savard and Polotskaia (2017) focused on implementing Davydov's ideas of introducing Grade 1–3 students to the analysis of word problems with additive structures from a relational perspective. In Norway, a research group led by N. Blank has adapted Russian textbooks by Arginskaya et al. (e.g., 2014) to the Norwegian Grades 1 to 4, currently implemented in more than 100 schools (Gjære & Blank, 2019).

This project, called Developmental Education in Mathematics (DEM), draws on the didactical system developed and experimented by Zankov in the 1950–70s. His didactical approach for elementary education has inspired today's Norwegian teachers to renew their practices by inviting students to solve difficult and challenging problems that develop a deeper understanding of mathematics and that nurture students' curiosity, engagement, intellectual autonomy, and overall capacity to learn (Zankov, 1975/1977; Melhus, 2015; Rennemo et al., 2018; Gjære & Blank, 2019). Teachers using Zankov's system in Norway have found their students more reflective and observant of their own mathematical activity and inclined to take on new challenges (Gjære & Blank, 2019).

Based on these initial and rather promising results, during the first author's research stay in New Brunswick, Canada, we selected a number of tasks from the Norwegian edition of Arginskaya's textbooks. Subsequently, we designed a set of lessons for Grade 6 students incorporating prealgebra tasks such as the analysis of numerical equalities, solving equations, and word problems.

In a preliminary study, we found that the lesson described here resulted in a range of momentary changes in the patterns of the students' mathematical noticing (Freiman et al., 2019). Building on this preliminary result, and in accordance with the principles of Zankov's mathematics lesson, our aim in this chapter is to investigate the teaching-learning conditions for coconstructing and expanding the Zone of Proximal Development (ZPD) that could have contributed to these changes.

The choice of this objective is based on Vygotskian ideas of analysis of the process of students' developmental potential based on increasing awareness of the complexity of mathematical concepts and the flexibility of their thought through active participation in a collective coconstruction of meaning. This process is complex, dynamic, and continuing.

Hence, regarding the development of students' understanding of a concept (e.g., the equal sign), the ability of the student to express (even correctly) their thoughts does not mean the end of the developmental path but rather its starting point for further development. This implies that from the researcher's point of view, an analysis of a concrete lesson does not aim to show what the student has or has not understood but rather to shed light on the main internal general developmental paths that the child could follow under the influence of systematic teaching (Vygotsky, 1933).

It is important to keep in mind that this small-scale study does not report on an implementation of Zankov's developmental system. While a single or a few lessons inspired by Zankov's system can represent a promising case that can be studied from a micro-developmental perspective, a full study of Zankov's didactical principles in practice would require a longitudinal study of a systematic implementation, such as the one underway in Norway (Gjære & Blank, 2019).

Walshaw (2017) argues that Vygotsky's work used in the context of mathematics education is "the best known" but at the same time seems to be "the least understood" (p. 307). According to her, its dialectical tradition and focus on societal relations and activity, "a complex inter-related unit in which social, cultural, historical, cognitive elements come into play" are important to "capture the richness of the learning experience" (p. 307). By introducing the ideas of Zankov, who himself was a close collaborator of Vygotsky, we hope to add to the understanding of how Vygotskian insight, especially regarding working within students' ZPD, can support teachers in promoting growth in students' mathematical thinking while taking into account the relational nature of their developmental potential (Walshaw, 2017).

THEORETICAL PERSPECTIVES OF DEVELOPMENTAL TEACHING BASED ON ZANKOV'S APPROACH

Zankov's system rests on Vygotsky's insight that education leads development and is constructed around five central principles aimed at providing students with optimal conditions for learning and development. Representing a holistic approach to teaching-learning, these principles form an interconnected whole, meaning that it does not make sense to try to implement only one or some of them (Zankov, 1975/1977):

1. Teaching at a high (optimal) level of difficulty (appropriate level of challenge)
2. A leading role for theoretical knowledge (focus on conceptual understanding, abstractions and generalizations)
3. Moving rapidly through the curriculum (learning something new in every lesson)
4. Students' reflection on their own learning process
5. Systematic development of every student in the classroom

Zankov's system operationalizes the concept of the ZPD for classroom teaching (Guseva & Solomonovich, 2017; Zankov, 1975/1977). According to Vygotsky (1933), teaching must address those of the students' abilities that are emerging but not yet matured, and not the abilities that have already matured.

This is the essence of "proximal" learning and of Zankov's principle of high and optimal difficulty (Zankov, 1975/1977). According to Walshaw (2017), this means introducing concepts that are, for the students, "on the cusp of emergence" (p. 303). In a similar vein, Radford and Roth (2011) proposed the notion of a students' horizon of conceptual potentialities, explained as "a phenomenon that without being clearly articulated emerges into the students' horizon of attention" (p. 234).

Learning in the ZPD results in the expansion of agency (the capacity to act), and the (ongoing) result of such learning is the process of development (Roth, 2020). The zone of proximal development is often contrasted with the zone of actual development, which is characterized by those problems the students can solve by themselves and where teaching-learning is not effective in promoting development (Zankov, 1975/1977).

Following Vygotsky's theoretical standpoint, Zankov saw collective activity, whether with adults or with peers, as being at the core of the ZPD:

> This level [the ZPD] is identified by noting the kinds of problems that the child is unable to cope with by himself, but can solve with the aid of grown-ups, in collective activity, or through imitation. But what a child can do in cooperation with someone else today, he will be able to do alone tomorrow. (Zankov, 1975/1977, p. 18)

While the ZPD has often been seen as an asymmetrical relationship between a "more knowledgeable other" and a learner, recent reworkings of the concept has formed a more symmetrical model (Roth, 2020). For instance, the role of more knowledgeable other is dynamic, with students often leading the teacher toward their educational needs (Graven & Lerman, 2014; Roth & Radford, 2010).

And, importantly, the construction of a ZPD is not only dependent on the teacher selecting appropriate learning activities, but also on the students' acceptance of the invitation to join in the activity (Roth, 2020) and their collective ethical commitment to achieve its objective, which Radford and Roth (2011) termed *togethering*. Recently, a study on mathematics teaching in early years showed in detail how a ZPD relied not only on the teacher, but also on the active contribution and commitment of the child to maintaining the teaching-learning activity and achieving its (shared) objective (Breive, 2020).

In essence, any ZPD is thus a *co*construction between the participants. This allows for an analytical perspective where one does not necessarily see the teacher acting at all times as "more knowledgeable other" and students acting at all times as knowledge-deficient learners; rather, one conceives of the participants as an "educational team," all together oriented toward the activity of teaching-learning (Radford, 2010). Espousing a similar view,

Zankov's system broke with the teacher-as-authority culture that dominated USSR classrooms at the time (Boguslavsky, 2015, Guseva & Solomonovich, 2017, Zankov, 1975/1977).

As mentioned by Roth and Radford (2010), the ZPD of each participant of the teaching-learning activity are not "zones of agreements but also of tensions, disagreements, misunderstandings, conflict, and subversion" (p. 306). Thus, the ZPD reflects, according to Vygotsky (1997), a certain "collision" between the perspective of the teacher and that of the student. In Zankov's (1975/1977) view, such contradictions should be present in the curriculum (tasks), to naturally surface during the teaching-learning activity, increasing its intensity.

It is through overcoming these differences students (as well as teachers!) learn and develop, expanding their agency (Roth, 2020). In order to accomplish this teachers and students must come together in joint teaching-learning activity, coconstructing a ZPD (Radford & Roth, 2011; Roth, 2020). A recent study confirms that practices based on sociocultural principles can build on *tensions* between teacher and student perspectives to increase student agency in the classroom (Brown, 2020).

Moreover, following the Soviet psychological tradition, Zankov did not separate emotions and cognition, but saw both as integral and important to students' learning. Thus, while he states that students must be challenged and exert effort in order to learn efficiently, he at the same time underscores the importance for their emotional well-being and interest in the learning activity to promote this effort (Zankov, 1975/1977). He mentioned intellectual enjoyment, satisfaction at solving a difficult problem and sympathy for others' views as important while also warning against negative emotions such as self-doubt or dissatisfaction over a classmate's mistake (Zankov, 1975/1977).

Guseva and Solomonivich (2017), discussing the specific psychological features of Zankov's system, stated that students are emotionally engaged and ready to contribute to the joint creative activity "only if he or she feels comfortable, when the relationships between the teacher and students are sincere, friendly, and warm" (p. 781). The importance of enjoyment for students to overcome mathematical challenges (*challenges-as-enjoyable*) was also documented by Freiman (2006) and Walshaw (2017). Thus, an analysis of the classroom conditions that promotes (or inhibits) the coconstruction of a ZPD must inevitably include emotional aspects.

A Developmental Approach to Algebraic Thinking

Zankov (1968/1973) saw developing students' observational skills as important to their general development. His system promotes analytical and thoughtful observation, prompting students to look and think, compare, describe, relate the parts and the whole, and to see a concept or phenomenon from various perspectives.

In this context, the development of algebraic thinking can be illustrated by two fundamentally different perspectives on numerical expressions such as $7 + 2$, noted by Freudenthal (1973): one can see the expression as an addition operation (I have seven, I obtain two more, the result is nine) or as a number-object (the number that is two more than seven), which is the algebraic perspective. This duality of mathematical expressions (which goes beyond arithmetic) has been explored further by Sfard (1991) and by Gray and Tall (1994), who suggested that the ability to shift flexibly between a procedural and a conceptual perspective is a characteristic of successful mathematical thinkers.

Molina et al. (2008) suggested one avenue for teachers to develop their students' conceptual perspective of the equal sign, namely, by asking students to determine whether various types of numerical equalities were *true* or *false*. They further suggested encouraging students to look and think before calculating, and to formulate questions that promote comparison and analysis (Molin et al., 2008).

In the Norwegian DEM textbooks (e.g., Arginskaya et al., 2016, see Figure 8.1) this type of tasks is embedded in a very particular didactical system that provides students with even more complex challenges, as the example from the second-grade textbook shows. In this task, students are not only asked to identify the equalities which are *false* but also suggest modifications that make them *true* for all values of the variables.

Figure 8.1

Example Task From the Grade 2 Book by Arginskaya et al. (2016) Showing a Focus on True/False and on Fixing False Equalities With Letters

118

a) Copy the false equalities.

$4 + x = 5 + x$	$y - 2 = y - 4$	$6 + a = a + 6$
$5 + k = k + 3$	$9 - e = 7 - e$	$c - 6 = 6 - c$

b) Make changes to the false equalities so that they become true.

c) Try to find different ways to solve the task.

Source: Translated From Norwegian by the First Author.

Note that Zankov's approach, as well as others suggested by Russian psychologists, use letters in the early grades, not in order to introduce formal symbolism or manipulations with letters but to make students investigate relationships, find patterns, and express them in a general form. Hence, algebraic thinking emerges not as a result of practice and drill exercises but from creative processes of coconstruction of meaning (Freiman, 2021).

Furthermore, reflecting the dual nature of mathematical expressions, the tasks in the textbooks explicitly target the students' awareness of a difference in meaning between a sum as additive relationship (e.g., the expression 7 + 2 which can be equivalent to another expression, like 2 + 7) and its numerical value (e.g., 9, as the result of an operation of addition). This is why students starting from the first grade are frequently asked to compare or analyze expressions without calculating their values, for instance, ordering sums in a descending or ascending order when one term is kept constant and the other varies (Figure 8.2).

This use of the word *sum* promotes an object-perspective of additive expressions. For instance, the task a) in Figure 8.2 treats sums as objects to be analyzed and not as procedures to be carried out, leading students to discover patterns and make conjectures. On the other hand, students are to "find the value of the sums" in order to check if their conjectures hold (task b), prompting self-reflection.

Figure 8.2

Example Task From Arginskaya Et Al. (2016) Showing How the Concept of a Sum is Studied Without Calculating its Value With Additional Thought-Provoking Follow-Up Questions

81

a) Without calculating, place the sums so that their values appear in ascending order.

36 + 4	35 + 4	37 + 4
39 + 4	34 + 4	38 + 4

b) Find the values of the sums.
Did you place them correctly? If not, find the mistakes.

c) Which other sums could fit into the pattern above?
Write them down and find the values.

d) Write some other sums that have a value equal to the biggest value that you found in a).

Source: Translated From Norwegian by the First Author.

This also exemplifies Arginskaya's approach of including thought-provoking questions. Hence, tasks c) and d) aim to deepen the students' engagement by asking them to create something by themselves. Task c) points to a possibility to extend the pattern beyond giving examples. Here, students apply a general rule they discovered and validated in tasks a) and b) (working further with a pattern), while adding a level of uncertainty for the students: What is the sought-after pattern? How to describe it?

These implicit questions produce discussion and raise the whole task to a higher level of difficulty. Task d) lets students play with expressions in another way (e.g., given a value, find some sums with that value), thus analyzing the pattern from a different perspective. Overall, the task exemplifies at least three important aspects of Zankov's approach to teaching mathematics in the ZPD, namely, providing students with an appropriate level of challenge, analyzing concepts from various perspectives, and increasing students' responsibility for the mathematical activity based on their curiosity and engagement.

PLANNING AND IMPLEMENTING A ZANKOV'S LESSON: OUR STUDY

In our analysis of the students' investigations, we were interested in identifying the teaching-learning conditions that might reveal their emerging developmental pathways and allow for the coconstruction of a ZPD. In particular, based on the students' work with numerical equalities, we analyzed momentary shifts in their perspectives in respect to the relational nature of the equal sign, shifts that could potentially lead to their expanded agency (Roth, 2020) regarding equations and the algebraic structure of word problems in the future.

We worked with a group of 20 Grade 6 students at one middle school in a rural area of Southern NB, Canada. These students followed a French immersion program, meaning that while many had English as their mother tongue, the language of instruction was mostly French. However, during our lesson, the students were free to speak French or English, reflecting the multilingual background of the research team.

It is also important to notice that the school is known for pioneering new learning spaces with embedded digital technologies. The group we worked with was a part of a recent pilot study on characteristics of learning spaces and computational thinking skills (Chiasson, 2019).

This included a classroom with mobile desks and chairs (with wheels) for the students, and two very large whiteboards covering two opposite walls. In addition, the students had access to small whiteboards and individual iPads. There was also an Interactive White Board and two Apple TV screens.

In the established work organization in this classroom, students could freely move around the classroom and use whiteboards (big and small) at their convenience. They could also share their iPad screen on one of the Apple TVs or Interactive White Boards making their work visible to their peers. At the time of our study, this type of setting was unique in the provincial school system.

All students were involved in the lessons, but research data were collected only from the 14 students for whom we were able to obtain parent's consent. Overall, we conducted three 90-minute-long lessons with this group, led by two researchers together with the classroom teacher, in which we tried several of Arginskaya's tasks (besides the equalities I–III below, the lessons included solving equations and word problems).

As we mentioned in the introduction of this chapter, at the beginning of the lesson, we wrote the following three equalities on one of the big white boards:

$$24 \times 214 = 12 \times 428 \quad \text{(I)}$$
$$325 + 427 = 529 + 225 \quad \text{(II)}$$
$$6 \times 99 = 600 - 5 \quad \text{(III)}$$

The first two equalities were taken from the Norwegian edition of Arginskaya's Grade 4 (2016) textbook, and we added the third one to increase the level of difficulty to better reflect the students' age and school level.

The classroom teacher identified these tasks as appropriate for Grade 6 students in connection to the Algebra strand of the provincial mathematics curriculum (New Brunswick Department of Education and Early Childhood Development [NBDE], 2010). The students were first asked to work on their own, and they were free to form small groups. During their work on the task, both researchers were asking the students to explain what they did, how and why. Whole-class discussions were also used at different moments of the lessons, allowing students to share their observations and reasoning with the whole class to further probe their mathematical thinking, and sometimes to try to provoke shifts in their mathematical perspectives without providing direct explanations.

Both authors were acting as teachers-researchers (we label them TR1 and TR2 to facilitate reading), TR1 led the activities and TR2 was video-recording whole class discussion. TR2 also interacted with students, asking them additional questions about what they were doing, why and how, without providing any further guidance. The classroom teacher (CT) was also interacting with some of students during small-group work and whole class discussions. We also interviewed him after the lessons. To denote students, we use acronyms of the type S*n* (Student *n*), for example, S7 is "Student 7."

RESULTS: INVESTIGATING THE PROCESS OF COCONSTRUCTION OF THE ZPD DURING THE LESSON

Working from an interpretative paradigm, we analyzed the data by looking for key moments when the students' mathematical perspectives shifted during their investigations of the numerical equalities. We then connected these moments to the process of coconstruction and expansion of a collective ZPD.

Students' Initial Engagement With the Tasks: Actual Zone of Development

Based on our initial reading of the Grade 6 curriculum documents, we expected students to use the ideas of relationship and balance to approach the equalities. However, to our surprise, all students in the classroom immediately started to calculate the values of each side of equality (I) using a variety of working modes: mental calculation, written calculation, or technology such as calculators.

This was documented through TR2's tour of the classroom with a handheld camera. For instance, a group of three students (S5, S9, and S14) decided to work collaboratively, sharing the workload of calculating the values of each side of the equality. One student decomposed the multiplication 24×214 as $(20 + 4) \times 214$ and then used the distributive property to calculate first 20×214 and then 4×214. Simultaneously, his teammate used a grid method to calculate 12×480 (see Figure 8.3).

This and other examples captured on video show a variety of arithmetic approaches using different methods of decomposing the factors to calculate the value of each side of the equality to determine whether the equalities were *true* or *false*. They also show how the students organized themselves in small groups, discussing and working collaboratively. From the developmental perspective, we draw two conclusions from these observations: first, that the students understood that the values on both sides of the equal sign must be the same for the equality to be true.

Second, they could solve the problem by means of calculations without assistance. These conclusions helped us to establish their *actual zone of development* (i.e., what students could do on their own). The next step was to try to initiate a discussion, inviting students to the coconstruction of a collective ZPD with the goal of potentially provoking shifts toward more relational perspectives in the analysis of the equalities.

Figure 8.3

The Group Of S5, S9, and S14 Began by Sharing the Work of Calculating Both Sides of Equality (I)

First Movement Toward Coconstructing the ZPD

By the end of the 6 minutes of the lesson, when most of the students still pursued calculations (see Figure 8.4), one student's voice, close to TR2 with the camera, said (in French): "vingt quatre moins douze c'est douze" … "… deux cent quatorze … " (meaning: 24 – 12 = 12; … 214 ….). We interpret his statement as a beginning movement toward a different mathematical perspective. The use of the word "minus" indicates to us that the students' perspective was not yet multiplicative ("half-double"), but it was no longer related to direct calculations on both sides of the equality either.

For the class as a whole, the relational perspective was not yet visible—it was below the students' *horizon of attention* (Radford & Roth, 2010). However, as the brief statement above shows, the "seed" of a different way to see the equality was present already at an early stage of the lesson. A few minutes later (around 7:30), as most students were still pursuing calculations, TR1 decided to call for a general discussion saying that "some of you are very busy calculating the numbers there, so has anyone figured out the first one, whether it's true or not?"

One student from the group of 3 (S5, S9, and S14) showed something to TR1 who said, "You have the first one. So, 24 × 214, that is 5136. OK, well done. So, you guys are in the process of finding out whether it is true or not." So, at this point, there was no indication by the teacher that there could be another way to "see" this equality.

Figure 8.4

A Student Doing Calculations on Her Digital Tablet for the First Equality

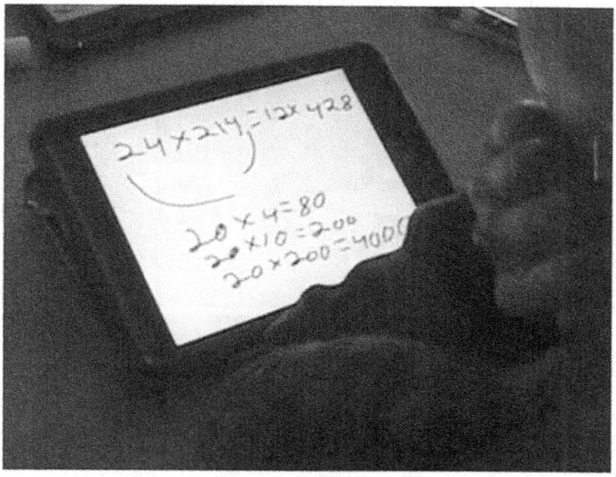

As students continued their work, another call came from TR1 (09:35). S10 (who might be the student whose voice was captured earlier on the camera) was encouraged by TR1 to share with the whole class: "Ici, c'est douze (*turns on microphone*) Ici, c'est moins douze ... ici ... et vingt-quatre, c'est le double." Translated into English, he said, "Here, it's twelve, here it's minus twelve ... here ... and twenty-four, that's the double." At this moment, the video-record shows the group of three students immediately reacting to S10's explanation. This key moment led to sudden and quite expressive gesturing and quiet but engaged discussion among S5, S9 and S14 (see Figure 8.5a and 8.5b).

Then TR1 said, "So, there's a pattern here. So maybe we don't have to calculate every single value. Do you guys see the pattern?" In response to this, S8 explained (in English) what she had noticed:

> S8 Twelve plus twelve equals 24 and 214 plus 214 equals 428, so they both equal the same.

Note that S8 said "plus," even though S10 said "double." Despite that the half-double relationship was not yet clearly expressed by S8 (she expressed additive reasoning), TR1 said, "Yes ... so it has something to do with halves and doubles." This remark pointed toward increased precision, not repeating S8's "plus" but instead emphasizing vocabulary related to multiplication.

Figure 8.5a and 8.5b

The Images Show How S14 Suddenly Noticed Something Based on S10's Explanation. The Camera Was Too Far Away to Pick Up Their Voices

However, TR1 did not explain the full reasoning process at that time. Instead, he encouraged the students to pursue their investigations further and think about what they had noticed: "So, maybe I have chosen these tasks so that you don't have to calculate everything. OK, so think about that a little bit."

Further Changes in Students' Perspectives as Result of "Togethering" in Small Groups

Around the 15:20 time mark, our attention was again drawn by the group of three students (S5, S9, and S14). Moving to equality (II), these students were playing around with the drawing of a balance scale with 0 on one side and 2 on the other side. Instead of putting the results of their calculations on both sides, they wrote the equality "0 = 2," focusing only on the difference of 2 (see Figure 8.6).

It is unclear how the students came to use the balance to represent their reasoning. However, the video-record shows a balance model in another episode when CT was discussing the same equality with another group of

Figure 8.6

Students Drew a Balance to Investigate the Second Equality

students. It could be that this model had been used by this class in the past to analyze the equality sign, especially since the use of balance models is recommended in the curriculum in the context of solving equations.

Yet, within our lesson, we interpret this as a clear representational shift toward a relational perspective compared to the students' work on the first equality event if they still based their conclusions on calculations of both sides of the equality. Moreover, the video-record shows that the students kept the same balance representation as their focus when moving on to equality (III). This series of three images (see Figures 8.7a, b, and c) shows first S14 (sitting, left) holding up a finger (see Figure 8.7a), indicating that there's a difference of 1 regarding equality (III), then S14 pointing to indicate that he wanted S9 (standing, middle, see Figure 8.7b) to erase the "2" on the right side of the balance (written during their work in equality II) and replace it with a "1" and finally, S9 accepting his suggestion and writing "1" (Figure 8.7c). S5 (see Figure 8.7c) was paying attention to S14's explanations. This small episode shows how the balance model served to focus the groups' *joint attention* (Radford & Roth, 2011) on the concept of difference (and not calculating values) during the discussion as they moved from the second to the third equality.

This episode also exemplifies a moment of *togethering* (Radford & Roth, 2011), where all three students were fully committed to the joint activity. They had distributed roles: S14 was leading and explaining, momentarily taking the role of more knowledgeable other in the group. S9 wrote S14's

Figure 8.7a

Students Using the Balance Model to Focus on Differences Between the Right and Left Sides of The Equal Sign, as They Moved From the 2nd to the 3rd Equality: Indicating a Difference of 1

Figure 8.7b

Students Using the Balance Model to Focus on Differences Between the Right and Left Sides of the Equal Sign, as They Moved From the Second to the Third Equality: Erase the "2"

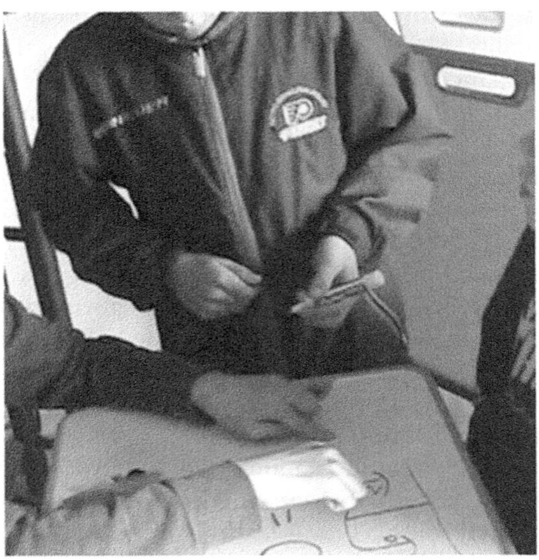

Figure 8.7c

Students Using the Balance Model to Focus on Differences Between the Right and Left Sides of the Equal Sign, as They Moved From the Second to the Third Equality: Accepting S9's Suggestion

suggestions, and S5 was paying attention—either monitoring the activity or trying to grasp S14's explanation (or both). Here, S14 took on the responsibility of being a leader of the discussion, and S9 and S5 trusted him in his leadership.

The group of S5, S9, and S14 showed great interest in the activity, at one point asking for more equalities (S9 said, "we would like a different one"). When they did not get one immediately, they instead returned to equality (I). This resulted in the rather unexpected 'doubling' (5136 + 5136) shown in Figure 8.8. Earlier, the group could have been able to reach a conclusion about the truth of the equality when they arrived at the identity 5136 = 5136 based on their calculations. However, later in the process, after having completed the second and third equalities, the students returned to the first one and started adding 5136 + 5136, circling 10272, the result of the addition (see Figure 8.8). Furthermore, when TR2 approached, S9 explained that they were working on the multiplication 5136 × 5136. A possible interpretation could be that this was a result of their (mis-)interpretation of the idea of "doubling." We interpret their willingness to revisit an already solved task as evidence of their curiosity and sustained effort in this activity. However, this finding also reveals a temporary limit for this group of students, an impasse in their developmental path. To overcome this impasse, another conflict would have to be introduced and resolved.

Figure 8.8

A Different Interpretation of "Doubling?" Students Adding 5136 And 5136

This was the last moment of this group recorded before the ensuing whole-class discussion, so their further path was not documented in this lesson.

Deepening Collective Awareness of a Novel Perspective

During the same phase of the lesson, we also we also found the work of student S2 interesing. She noticed a relationship between the terms on each side of equality (II), showing that there is a difference of 102 between 427 and 529 and a difference of 100 between 325 and 225. Moreover, her explanation shows that she could also use this relationship to suggest a way to modify the equality to make it true:

S2 (*exclaims*) OK, I found it, I found it, I think I found it!

TR2 OK, so what did you find?

S2 OK, so … this, to make this, you have to add one hundred, OK, (*points at 427, then at 529*) and this, to make this, you have to add one hundred (*points at 325, then at 225.*)

TR2 Right.

S2 So, you've got to change the numbers over, changing this to seven (*points at the units place of 529*), so that both are one hundred (*points at 427 and 529*) and one hundred (points at 325 and 225). So, then they're both one hundred (*gestures at the whole equality*).

Here, we note her tone of voice, which in our interpretation expressed enthusiasm when she made her discovery (a "Eureka" moment), that is, when the relationship became visible on her horizon of attention. We therefore conclude that S2 experienced the task as challenge-as-enjoyable (Walshaw, 2017).

Nevertheless, while she compared the terms by focusing on the relationships between them, the language she used to describe the differences of 100 reflects a process of calculation: "you have to add one hundred" (to get from 225 to 325) instead of one of balancing (keeping the same difference of 100 on both sides). Hence, this mixed language based on calculations and on differences to describe the structure of the equality in terms of numerical relationships suggests a tension between the two perspectives (Brown, 2020), not fully resolved at this moment.

While discussing the second equality with the whole class afterward (around 20:00), we tried to bring the students' attention to the possibility of finding different ways of changing the numbers to make the equality *true*. Hence, S2 shared her method of modifying the second equality (described above) with the class. TR2 then prompted the class to find a different way to make the equality true. In response to this call, another student (S3) came up with the following explanation, in which the focus is on the difference of 102 between 327 and 225, trying to fix the balance on the other side of the equality:

(Written on the whiteboard: 325 + 427 = 529 + 225)

S3 Well, I was trying to think ... you could add 2 to the 325 ... maybe?

TR1 So, you want to add 2 to the 325. Why would that make it equal? (*TR1 points at 325 on the board*).

S3 Hmm, 102. So that if you added 2 to that, then you equalize the 225.

TR1 Yeah, so we compare again. So, let's say this is 327. Can I change it? (*TR1 erases the digit 5 in 325 and puts down 7 instead.*) There's 327 (*points at 327*) and here is 225 (*points at 225*). And you said equalize or compare?

S3 102.

TR1 It's 102, the difference. This is 102 bigger, right? (*TR1 points at 325*).

S3 And then 529 is 102 bigger.

TR1 And then this one (*points at 529*) is 102 bigger than that one (*points at 427*).

The new aspect above is that S3 had noticed another numerical relationship to use to make the equality true ("equalizing"), using a difference of 102 instead of 100. This suggests a expansion of S3's perspective, from looking for a single relationship to use to noticing that there are several ways to look at an equality, offering different paths to a solution and seeing the problem from various sides.

Equality (III) allowed some students to adopt a relational perspective based on the distributive property. As with the equalities (I) and (II), the students were initially engaged in calculating 6 × 99 through various procedures before the TRs and the CT hinted that 99 could be close to a "neat" number. During the ensuing whole-class discussion, two students (S3, S4) then shared a "new" perspective with the class (see Figure 8.9). In a lively discussion between the two students, even as they were presenting their solution to the class, S3 suggested changing 99 to 100 in the following way: 100 × 6 – 6 = 600 – 6. S4, on the other hand, suggested keeping 99 on the left side of the equality and only changing 5 to 6 on the right side (99 × 6 = 600 – 6) to make the equality true:

(Written on the board: 6 × 99 = 600 – 5)

S3 You add one to this to make 100 (*erases 99 and writes 100*). Then, you minus the 6. You minus it because you added one, and one times six is six (*writes – 6 after 100*).

S4 (*Points at 6 × 100 – 6*).

S3 Which is five hundred and ninety ... four! So, if you change this into a six instead of a five, you're going to have five hundred and ninety-four (*points at the 5 in the right-side expression*).

(They confer quietly for a few seconds).

S4 You can also, instead of changing this (*points at 100 – 6*), you can keep the 99, and instead of minus five here (*points at the 5 on the right side*), you put minus six, so like ... (*erases 100 – 6 and puts 99 back*) and this to a 6 (*erases 5 on the right side and writes 6 instead*).

The key shift in perspective occurring here was one from using multiplication procedures to calculate 6 × 99 to noticing that 99 is close to 100 and exploiting this using the distributive property. In the equalities (I) and (II), the numerical relationships the students analyzed were between numbers present in the equality. In equality (III), however, the relationship was between the expressions 6 × 99 and 6 × (100 – 1), and the number 100 was

Figure 8.9

S3 and S4 Engaged in Discussion on How to Interpret Equality III While Presenting to the Class

not present in the original equality—the students had to "see" something that was "hidden."

Students Taking Control of the Lesson, "Creative Polyphony"

Throughout the lesson, we noted the students' willingness to share ideas with each other. In general, students often share ideas when asked or encouraged by the teacher. However, in this lesson, the students also shared their work with us and with each other spontaneously on many occasions. One means by which this happened, was through the TV screens that were placed in the front and the back of the classroom. All students had access to cast from their iPads to these screens, anonymously and at any time. One such example is given below (see Figure 8.10), showing an (unknown) students' proposed solution to equality (II).

We also note the "smiley" drawn to the left, perhaps indicating that the student was enjoying the activity. The fact that sharing in this way was anonymous could plausibly lower students' emotional risk when sharing solutions. We recorded at least four different students sharing their screens at various points in the lesson.

Another way some students made their work public was by choosing to work on the large whiteboard that covered a whole wall in the back of the room, such as S7 (left) and S6 (right) shown below (see Figure 8.11). The work of S6 and S7 in this picture was preceded by the start of the ensuing whole class discussion, when TR1 reminded students about the half-double

Figure 8.10

TV Screens in the Classroom Allowed Students to Screencast Anonymously From Their Individual Digital Devices

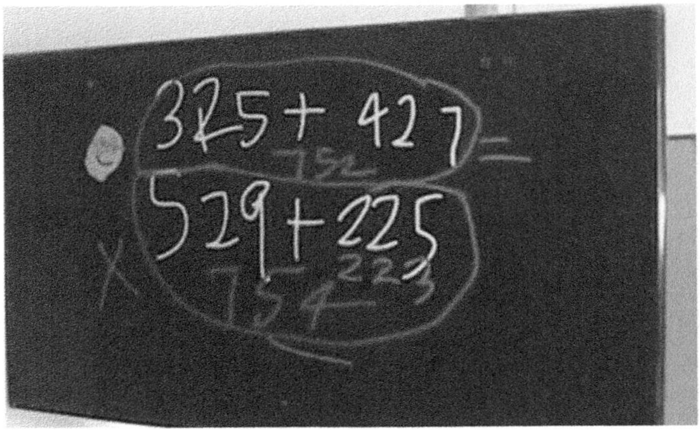

relationship in equality (I). Perhaps this prompted the two students, who had already completed equalities (II) and (III), to go back to equality (I) and to investigate it further.

In the video, S6 is seen working by himself (eventually with the help of S7) on a whiteboard section of the wall, continuing even as the whole-class discussion was moving on to equalities II and III, writing the following list of equalities:

$$24 \times 214 = 12 \times 428$$

$$48 \times 428 = 24 \times 856$$

$$96 \times 856 = 48 \times 1712$$

$$192 \times 1712 = 96 \times 3424$$

(Continuing five or six more lines.)

Here is his explanation to the class afterward:

> S6 (....) doubling all the numbers over and over and over. And now I'm here (pointing at the last line he has written).

Figure 8.11

S6 and S7 Making Their Joint Work Public by Writing on a Large Blackboard on a Wall Section in the Back of the Classroom

In fact, S6 had produced a pattern of $a \times b = a/2 \times 2b$, albeit not yet expressed in a general symbolic form. The way he kept on suggests he noticed that this pattern could be extended indefinitely. This is supported by him saying "over and over and over." S6 investigated and expanded the equality independently, staying within the half-double relationship for some time through his own activity. For S6, the general form of the pattern had just emerged on his horizon of (possible) attention. By doing this, he had left the initial true/false-problem and instead engaged in an ongoing creative process of calculating the doubles of each factor, continuing the pattern he had discovered and sharing it with the class eventually extending the collective ZPD.

LESSONS FROM "ZANKOV'S LESSON": THE ROLES OF THE TEACHERS, THE STUDENTS, AND THE CLASSROOM IN PROVIDING THE CONDITIONS FOR COCONSTRUCTING A ZPD

Overall, we identified and described evidence of some (momentary) shifts in the students' mathematical perspectives. The data indicates not only changes in perspective, but also a concurrent change in the nature of the mathematical activity: At first, the goal of the task, as perceived by all students, was to determine if the values on both sides of the equal sign were the same, using known procedures. However, when alerted to the possibility of solving the task by analyzing, the goal was transformed for some of the students: The task changed into finding some numerical relationship to use to justify their answer. The students came together to coconstruct a ZPD (togethering) and were able to produce a range of novel mathematical ideas.

Our data also suggests that this change in perspective is not of a binary nature, that is, it is not like flicking a switch between one perspective and the other (process and object). While some students were able to expand their perspective during this lesson to include analyzing numerical relationships, it was clear that they were struggling to explain it in a clear way and even to make sense of it.

This was evidenced by students describing a 'half-double' relationship in an additive way (e.g., when S8's responded to S10 saying "double"); by students comparing differences between terms using a language of operation (S2: "you add 100 ..."); and by some unexpected interpretations of the idea of "doubling" (e.g., the group of three students adding 5136 + 5136, or S6 continuing his list of equalities, constantly doubling all factors). The new perspective had entered the students' horizon of attention (Radford & Roth, 2011) but they could not yet see it clearly.

From a developmental theoretical perspective, this is to be expected and even encouraged, since students' mathematical development can be "non-linear" in nature and follow diverse paths (Coles & Sinclair, 2018). As emphasized by Vygotsky (1933), each lesson is not a finished product, but the *starting point* for the next step in the learning process. The tension experienced by the students in this lesson thus became the motivation for us to revisit the topic of true/false equalities in our next lesson, to continue expanding the collective ZPD.

While the analysis of the second lesson is beyond the scope of this chapter, we were able to document more lasting changes in some students' reasoning. Below, we return to the first lesson to investigate in more depth how the roles of the teachers and students could plausibly have been conducive to this change.

Actions by the Participants, Relevant to the Coconstruction of the ZPD

One important characteristic of the lesson was that TRs and the CT gave the students free rein but at the same time provided the students with indirect hints. For example, around the 03:45 mark in the video, shortly after the introduction of the task, CT said loudly to the whole class, "I see some magic!"

Furthermore, after allowing some students to share what they have noticed about the first equality, TR1 said (11:25), "Maybe I've chosen these tasks so that you don't have to calculate everything. OK, so think about that a little bit." Generally, the TRs refrained from providing the students with direct explanations, instead inviting the students to the challenge of reflecting and making sense of the new perspective. CT noted this general approach during the interview after the lessons:

> CT You had a fresh way of bringing it to the class and that was fun. I mean, your questions were interesting, they were not drills, and they were not just a repetition of the same thing. They were really neat to see, and then when you questioned somebody, it was almost as if you put a puzzle to somebody. That part is fun. Some grab on to that, especially the kids that love that kind of challenge.

The hints and puzzles that the TRs provided for the students were aimed the emotional aspect of mathematical activity. By being inviting and trying to create a 'mystery' for the students we wanted to build interest and curiosity as a motivating force for their investigations, which Zankov (1975/1977) saw as important and which Walshaw (2017) termed *challenge-as-enjoyable*.

The challenge-as-enjoyable was noted by the students as well. In fact, in the post-lesson interviews with six pairs of students, some mentioned the true/false tasks described in this study as the most enjoyable of all the problems we gave them:

> TR2 Was there any particular task that you remember, could be hard or easy, that you really liked?
>
> S1 The equals? That was really *fun*.
>
> S8 Yeah, that one.
>
> TR2 Why did you like that?
>
> S8 Honestly, I don't know, I just *really liked* doing it.

Several other students also mentioned key words such as challenge and struggle as an enjoyable part of the lessons, even reflecting on the importance of challenge for their own mathematical development:

> TR1 So, I just want to ask you about the lessons. How were they? (…)
>
> S10 They were pretty good, because they were *challenging*. And once you figured it out, it seems pretty easy.

TR1 So how was it compared to the usual lessons?

S11 This is usual lessons (*gives thumbs down*). This is ... (*gives thumbs up*)

TR1 So what's different?

S11 There's more *challenge*. It's algebra, which we don't do very often. It's not IT.

S10 We can do it on paper. It's more complex, so it's challenging your brain. So once you figure it out you can move on to another level of math.

Here the Zankov's principle of making students aware of their own learning activity and progress (Zankov, 1975/1977) might be helpful to interpret student's self-reflection on the experience with our tasks. S11 referred to IT (information technology), echoed by S10 ("We can do it on paper. It's more complex ..."), perhaps indicating that they sensed a lack of complexity and challenge in their routine work with computer-supported applications with more traditional content (exercises for practicing skills). We did not analyze these and therefore cannot assess the students' statements about this topic. However, we do note that "Zankov's lessons" prompted the students to reflect on different approaches to teaching elementary mathematics and to notice differences with more traditional tasks.

We also note that students' sharing of ideas always was met respectfully in this class. The TRs and the CT took care to acknowledge and encourage the thinking processes of all students, including those who might have remained with a rather stable calculation perspective, or those who might had made a mistake. This attitude was shared by the students, lowering the emotional risk of expressing mathematical ideas and providing a safe atmosphere that is conducive to learning where the *contribution of every student is valuable* (Guseva & Solomonovich, 2017; Zankov, 1975/1977).

Furthermore, by allowing the students free choice of method (and, as it turned out, task interpretation), we wanted to promote their co-ownership of the problem-solving process, empowering the students. These actions by the teachers therefore also relate to the concept of togethering, as they promote the students' coresponsibility for moving the lesson forward.

At the same time, for the students to accept this invitation to take responsibility, some trust must already be in place so that students feel safe in participating. Togethering is a mutual ethical commitment which is the basis for joint activity, but since joint activity is constantly ongoing and emerging, togethering can be seen as a result of joint activity (Breive, 2020), revealing the dialectical "chicken-and-egg" nature of the process of establishing a ZPD.

Properties of the Classroom, Relevant to the Coconstruction of the ZPD

First, students were allowed and encouraged to talk to each other and share ideas. The result was that they immediately formed small impromptu groups or pairs working together. Moreover, as the lesson progressed, the groups overlapped or changed as students began moving around, talking to other students. This was captured on video, for example, when S2 near the end of the group work session had moved to the back of the room, where the pairs of S3 and S4, and S6 and S7 were working.

This dynamic interaction was made possible by the physical properties of the classroom. This particular room was equipped with a type of chair-and-desks on wheels, allowing students to roll around the classroom while bringing their workstation with them. Furthermore, we earlier noted the TV screens (see Figure 8.9) allowing for anonymous sharing, and the large whiteboard areas on the walls (see Figure 8.10), tools allowing students to share written work.

Also, the classroom was equipped with wireless microphones that students and teachers could use when addressing the whole class, making it easier to capture their joint attention. Using these, the TRs had the option of strategically selecting and encouraging certain students who had noticed mathematically important ideas to share them with the class, such as when S10 explained that 24 is double of 12. In sum, we find that both the physical organization of this particular classroom and the relatively safe atmosphere where children could share their ideas freely without emotional risk (Chiasson, 2019) allowed and encouraged the collective mathematical activity, thus laying the foundation for the process of togethering and eventually the expansion of the collective ZPD.

CONCLUSIONS

In this study, we have presented a unique case based on Zankov's (1975/1977) educational system, where two teacher-researchers invited a Canadian (NB) sixth grade group to analyze numerical equalities inspired by the Norwegian version of Russian textbooks. With this study, we want to bring to light how Zankov's ideas can contribute to current classroom practices by providing teachers and students with better conditions in which to realize coconstruction of the ZPD.

In the analyzed lesson, we first found that the students considered the task to be manageable, but their task interpretation was not the same as the TRs'. While we expected the students to look for numerical patterns or relationships to use in order to determine whether the equalities were *true* or *false*, they interpreted the task as calculating both sides of the equal sign and comparing the results, despite TR1 using formulations such as "look at them, investigate, and think about them a bit."

This revealed a conflict (Vygotsky, 1997) or tension (Brown, 2020) between the perspective of the TRs and that of the students. Our case study thus provides some details on how such tensions can be brought to the surface. The structure of the equalities we used in the lessons contributed to revealing tensions, since they were deliberately created to promote analysis of structure and not calculation.

The characteristics of a task or concrete materials can provide the students with a problem that not only challenges them, but also helps them to produce certain new ideas, acting as an "absent more knowledgeable other" (Abtahi, 2017). Both Zankov (1997/1965) and Arginskaya et al. (2014) produced mathematics textbooks with thought-revealing tasks designed with systematic general development of every child as a goal, helping students and teachers move forward, eventually expanding the ZPD.

Second, we found that some students showed changes in their mathematical perspectives. These potential cognitive gains were accompanied by increased intensity in the students' emotional reactions, evidenced by their gestures, face expressions, tone of voice and eager discussions. Not only did the socioemotional intensity of the lesson increase, but also the cognitive intensity brought on by tensions between differing perspectives.

This resulted in the transformation of the nature of the task itself, signifying a shift of attention from calculations based on operations to relationships and patterns, with students trying to make sense of them. This finding demonstrates the integral nature of cognition and emotions noted by Zankov (1975/1977), showing the importance of challenge-as-enjoyable (Walshaw, 2017) for moving learning forward.

A third important finding, related to the notion of *agency* (Roth, 2020), was how students gradually took control of the teaching-learning activity. At the beginning, students seemed to follow instructions (as they interpreted them) from TR1. However, as the lesson progressed, and noticeably after the key exchange happening at the 10-minute mark, we noted spontaneous actions from the students that were outside the control of the TRs.

Examples include S6 and S7, who extended the pattern of equalities; the group of S5, S9 and S14, who returned to equality (I) (without anyone asking them to) and started experimenting; students bringing in the balance model; students sharing their ideas on screens; and students moving around in the classroom to discuss with their classmates. We termed this situation a "creative polyphony," where various small groups of students were simultaneously working in several directions within the same classroom.

We located the key moment that preceded all these changes to around 10 minutes into the lesson, when S10 explained that 24 is the double of 12, with S8 and TR1 following up. It is noticeable that the shift was not explicitly directed by the TRs, who did not provide any direct explanations but instead posed more questions to challenge the students. This teacher-supported ambiguity resulted in a polyphony of activity in the class, not driven by the TRs' directions, but by the curiosity of those students who wanted to make sense of the new perspective that had emerged on their horizon of attention (Radford & Roth, 2011).

Our findings suggest some potential benefits of a "creative polyphony" of students' voices and actions, even if at some moments of the lesson, the student-led mathematics classroom tended to appear "chaotic." Teachers need to balance and harness such chaos in order to optimize conditions for students' learning and development (e.g., Russo, 2019).

Moreover, our analysis shed light on the notion of *challenge* as central to Zankov's (1975/1977) system, not only in the sense that the task should be appropriately difficult, but also in the sense that providing students with thought-revealing questions creates a type of "mystery" that can increase students' intellectual enjoyment and further motivate their engagement in the learning activity. Moreover, Zankov (1999) advised allowing students room to think for themselves and make their own mistakes, seeing a mistake as a steppingstone to a solution of the problem in a risk-free classroom atmosphere.

Overall, the students seemed to respond to our invitation with engagement and enthusiasm showing curiosity and sharing their discoveries and reasoning, thus offering an invitation themselves: They invited us into *their own world* of mathematical thinking. This symmetrical cognitive and socio-emotional relationship between teachers and students further illustrates that the ZPD is always *co*constructed.

The ZPD can only arise through interaction in the moment of the teaching-learning (e.g., Breive, 2020), showing the difficulty of planning a lesson ahead in explicit detail as it requires much flexibility from teachers. However, the teacher can employ certain strategies to try to be tuned to the students' thinking, so that s/he makes coconstruction of the ZPD easier: listening, being open to ideas, giving students time to think, allowing them the opportunity to share, organizing a classroom that facilitates free communication, and getting to know the students' thought patterns over the time (Zankov, 1999).

In this respect, our case study seems to demonstrate potential gains students could make when teachers are open and sensitive to their multiple developmental pathways. This work includes providing high-level thought-provoking tasks, an appropriate level of challenge, targeting conceptual understanding and students' reflection on their own learning, as well as recognizing everyone's creativity and various perspectives as a rich foundation for coconstructing a collective ZPD.

Despite decades of various reforms in mathematics education, there is still a need for teaching practices where all students have an opportunity to participate and be heard, and to develop according to their own potentials and interests (Schnell & Prediger, 2016). While we suggest caution in generalizing based on this single and time-limited case, we nevertheless draw attention on the importance of further expanding implementations of developmental teaching practices that are based on earlier works of educational innovators but still provide novel perspectives in line with today's debates and reforms.

REFERENCES

Abtahi, Y. (2017). The more knowledgeable other: A necessity in the zone of proximal development? *For the Learning of Mathematics, 37*(1), 35–39.

Arginskaya, I., Benenson, E., Itina, L., Kormishina, S., Blank, N., & Melhus, K. (2014). *Matematikk Grunnbok 1A*. Barentsforlag.

Arginskaya, I., Ivanovskaya, E., Kormishina, S., Blank, N., & Melhus, K. (2016). *Matematikk Grunnbok 2A*. Barentsforlag.

Breive, S. (2020). Student-teacher dialectic in the co-creation of a zone of proximal development: an example from kindergarten mathematics. *European Early Childhood Education Research Journal, 28*(3), 413–423. https://doi.org/10.1080/1350293X.2020.1755498

Brown, R. (2020). Re-conceptualizing the development of agency in the school mathematics classroom. *Theory Into Practice 59*(2), 139–149. https://doi.org/10.1080/00405841.2019.1702394

Boguslavsky, M. L. (2015). Vital pedagogy of Leonid Zankov. *Russian-American Education Forum: An Online Journal, 7*(3). www.rus-ameeduforum.com/content/en/?&iid=23

Chiasson (2019). *The relationship between learning spaces and the development of computational thinking skill* [Paper presentation]. The 2019 annual meeting of the American Educational Research Association. Retrieved on October 12th, 2020, from the AERA Online Paper Repository. https://doi.org/10.302/1440741.

Coles, A., & Sinclair, N. (2018). Re-thinking 'normal' development in the early learning of number. *Journal of Numerical Cognition 4*(1), 136–158. https://doi.org/10.5964/jnc.v4i1.101

Freiman, V. (2006). Problems to discover and to boost mathematical talent in early grades: a challenging situations approach. *The Montana Mathematics Enthusiast, 3*(1), 51–75.

Freiman, V., & Fellus, O. O. (2021). Closing the gap on the map: Davydov's contribution to current early algebra discourse in light of the 1960s Soviet debates over word-problem solving. *Educational Studies in Mathematics, 106*(1), 343–361. https://doi.org/10.1007/s10649-020-09989-6

Freiman. V., Gjære, Å. L., & Blank, N. (2019, April 5–9). *Shifts in primary school students' mathematical noticing when analyzing numerical equalities* [Poster presentation]. The 2019 AERA annual meeting, Toronto, Canada.

Freudenthal, H. (1973). *Mathematics as an educational task*. D. Reidel.

Gjære, Å. L., & Blank, N. (2019). Teaching mathematics developmentally: Experiences from Norway. *For the Learning of Mathematics, 39*(3), 28–33.

Graven, M., & Lerman, S. (2014). Counting in threes: Lila's amazing discovery. *For the Learning of Mathematics, 34*(1), 29–31.

Gray, E. M., & Tall, D. O. (1994). Duality, ambiguity, and flexibility: A "proceptual" view of simple arithmetic. *Journal for Research in Mathematics Education, 25*(2), 116–140. https://doi.org/10.2307/749505

Guseva, L., & Solomonovich, M. (2017). Implementing the zone of proximal development: From the pedagogical experiment to the developmental education system of Leonid Zankov. *International Electronic Journal of Elementary Education, 9*(4), 775–785.

Kieran C. (2018). Seeking, using, and expressing structure in numbers and numerical operations: A fundamental path to developing early algebraic thinking. In C. Kieran (Ed.), *Teaching and learning algebraic thinking with 5- to 12-year-olds. The global evolution of an emerging field of research and practice* (pp. 79–105). Springer. https://doi.org/10.1007/978-3-319-68351-5_4

Melhus, K. (2015). Å stimulere barns evne til å tenke [Stimulating children's ability to think] *Tangenten – tidsskrift for matematikkundervisning 26*(2), 13–16.

Molina, M., & Ambrose, R. (2008). From an operational to a relational conception of the equal sign. Thirds graders' developing algebraic thinking. *Focus on Learning Problems in Mathematics, 30*(1), 61–80.

Molina, M., Castro, E., & Mason, J. (2008). Elementary school students' approaches to solving true/false number sentences. *PNA, 2*(2), 75–86.

New Brunswick Department of Education and Early Childhood Development. (2010). *New Brunswick Mathematics Grade 6 Curriculum Guide*. https://www2.gnb.ca/content/dam/gnb/Departments/ed/pdf/K12/curric/Math/Math-Grade6.pdf

Pang, J., & Kim, J. (2018). Characteristics of Korean students' early algebraic thinking: A generalized arithmetic perspective. In C. Kieran (Ed.), *Teaching and learning algebraic thinking with 5- to 12-year-olds* (pp. 141–165). Springer. https://doi.org/10.1007/978-3-319-68351-5_6

Radford, L. (2010). The anthropological turn in mathematics education and its implication on the meaning of mathematical activity and classroom practice. *Acta Didactica Universitatis Comenianae Mathematics, 10,* 103–120.

Radford, L., & Roth, W. M. (2011). Intercorporeality and ethical commitment: An activity perspective on classroom interaction. *Educational Studies in Mathematics*, 77(2–3), 227–245. https://doi.org/10.1007/s 10649-0 10-9282-1

Rennemo, M. G., Søvik, W. L., & Meberg, L. K. O. (2018). Utviklende matematikklæring [Developmental learning of mathematics]. *Tangenten – tidsskrift for matematikkundervisning, 29*(1), 15–20.

Roth, W. M. (2020). Zone of proximal development in mathematics education. *Encyclopedia of Mathematics Education*, 913–916. Springer. https://doi.org/10.1007/978-94-007-4978-8_156

Roth, W. M., & Radford, L. (2010) Re/thinking the zone of proximal development (symmetrically). *Mind, Culture, and Activity, 17*(4), 299–307. https://doi.org/10.1080/10749031003775038

Russo, J. A. (2019). Walking the line between order and chaos: A teacher-researcher's reflection on teaching mathematics with challenging tasks in primary classrooms. *International Journal of Innovation in Science and Mathematics Education (formerly CAL-laborate International), 27*(3), 14–24. https://openjournals.library.sydney.edu.au/index.php/CAL/article/view/13210

Savard, A., & Polotskaia, E. (2017). Who's wrong? Tasks fostering understanding of mathematical relationships in word problems in elementary students. *ZDM, 49*(6), 823–833.

Schnell, S., & Prediger, S. (2017). Mathematics enrichment for all—Noticing and enhancing mathematical potentials of underprivileged students as an issue of equity. *EURASIA Journal of Mathematics, Science and Technology Education, 13*(1), 143–165. https://doi.org/10.12973/eurasia.2017.00609a

Sfard, A. (1991). On the dual nature of mathematical conceptions: Reflections on processes and objects as different sides of the same coin. *Educational Studies in Mathematics, 22*(1), 1–36. https://doi.org/10.1007/BF00302715

Vygotsky, L. S. (1933). O pedologicheskom analyze pedagogicheskogo processa [On pedological analysis of pedagogical process]. In L. S. Vygotsky (Ed.), *Pedagogicheskaya psychologiya* [Pedagogical psychology] (pp. 373–392). Pedagogika.

Vygotsky, L. S. (1997). Development of arithmetic operations. In R. W. Rieber (Ed.), *The collected works of L. S. Vygotsky: The history of the development of higher mental functions* (pp. 149–152). Plenum Press.

Walshaw, M. (2017). Understanding mathematical development through Vygotsky. *Research in Mathematics Education, 19*(3), 293–309. https://doi.org/10.1080/14794802.2017.1379728

Zankov, L. V. (1973). *Didaktik und Leben: Beitrage zu einer neuen Didaktik* [Didactics and life: Contributions to a new didactic] (B. Schiff, German, Trans.). Hermann Scroedel Verlag. (Original work published 1968)

Zankov, L. V. (1977). *Teaching and development: A Soviet investigation* (A. Schultz, Trans.). M. E. Scharpe. (Original work published 1975)

Zankov, L. V. (1999). Besedy s uchiteljami [Conversations with teachers]. In L. V. Zankov (Ed.), *Selected pedagogical works* (pp. 470–593). Dom pedagogiki.

Zankov, L. V., & Zankov, V. V. (1997). *Uchebnik matematiki dlya 1 klassa* [Mathematics textbook for Grade 1]. Moskva, Dom pedagogiki. (Original work published 1965)

CHAPTER 9

ENGAGING ELEMENTARY PRESERVICE TEACHERS WITH HIGH LEVERAGE MATHEMATICAL PRACTICES

Christopher Nazelli, Deborah Zopf, and S. Asil Özgün-Koca

INTRODUCTION

The Joint Position Statement of the Mathematical Association of America and the National Council of Teachers of Mathematics on Teaching Calculus holds that:

> The college curriculum should offer students an experience that is new and engaging, broadening their understanding of the world of mathematics while strengthening their mastery of tools that they will need if they choose to pursue a mathematically intensive discipline. (n.d., p. 1)

This is especially true for students who wish to become teachers of mathematics. Further, the American Mathematical Society cites Ken Bain's (2004) work indicating that instructors who foster classrooms in which "people learn by confronting intriguing, beautiful, or important problems, authentic tasks that challenge them to grapple with ideas, rethink their assumptions, and examine their mental models of reality" (p. 1) tend to be highly effective. Stigler and Hiebert (2004), in their international

Mathematics as the Science of Patterns:
Making the Invisible Visible to Students Through Teaching, pp. 185–203

comparisons, found that *how* a class interacts about mathematics can be more important even than *what* mathematics they interact with. Unfortunately, making meaning while doing mathematics is rare in United States classrooms, while memorization and procedure-following are more commonplace (Moses & Cobb, 2001). Keeping these suggestions in mind, in 2015, members of Wayne State University's (WSU) College of Education and Henry Ford College (HFC) embarked on a journey to develop high-quality mathematics courses for preservice teachers (PSTs). Our work focused on the redesign of two Mathematics for Elementary Teachers courses at WSU and equivalent courses at HFC. We intended to redesign the courses to reflect the mathematics—content and practices—important for preservice elementary and middle school teachers as well as pedagogical strategies that engage these learners of mathematical knowledge for teaching (Ball et al., 2008). Therefore, we aimed to design courses that would provide challenging mathematical tasks intended to elicit deep understanding of fundamental mathematics, develop mathematical practices appropriate to the discipline, and foster written and oral communication skills important for the work of teaching.

THE APEX CYCLE—PEDAGOGY TO DEVELOP MATHEMATICAL KNOWLEDGE FOR TEACHING

As our work with PSTs evolved, we noticed that it took on an interesting structure which we call the APEX Cycle (see Figure 9.1). When we launched an activity, PSTs engaged in an *Analysis* phase during which they explored several questions focused on a particular topic. Often the questions asked the PSTs to use multiple approaches to explore a mathematical topic or procedure. They used representations and manipulatives that allowed them to see the mathematics through different lenses.

After a period of exploration and independent thinking, we shifted to the *Presentation* phase, during which PSTs presented their thinking to the whole class. This phase opened up many learning opportunities. First, the presenters had to communicate their mathematical work using correct mathematical language. They could no longer feel comfortable "knowing it in my head" but had to put into words mathematical ideas. They worked as a group to develop correct language to communicate mathematical ideas. Second, as a class, the PSTs responded to the presentation. They developed respectful ways to comment on the ideas of their peers, to recommend revisions, and to develop a common language with which to discuss their work.

With the extensive conversation complete, we transitioned to the *Explanation* phase during which PSTs wrote the solutions and explanations of their understanding. They were guided by several prompts: write to a peer, include important conceptual details, justify your thinking, and explain not only how but why. PSTs developed their abilities to think through and write about procedures with attention to their conceptual underpinnings. This deeper understanding allowed the PSTs to see the standard algorithms as procedures that *could* be done rather than *must* be done.

Then the PSTs *eXchanged* journals across institutions. In this exchange, PSTs reviewed another's journal examining it for communication of mathematical ideas. In this review, PSTs commented on their peer's approach to a problem, explanation of process, and development of mathematical ideas. We asked PSTs to review a peer's journal by responding to specific prompts: (1) What is the mathematics in the work?, (2) What did the author do well?, (3) How can the work be improved?, and iv) Write a friendly note to the author.

Even though the APEX cycle in Figure 9.1 might look like a linear process, it is repeated for different tasks and each iteration builds on students' mathematical thinking and their readiness for the next cycle. All the components of the APEX and cyclic nature are supported with multiple forms of communication, the interplay between the conceptual and procedural understanding, and the freedom from prescribed ways of doing mathematics. Negotiating each other's individual understanding during the analysis stage via group work helps students to form their collective mathematical understanding. When presenting their work to the whole class during the presentation stage, that collective understanding is made public. During each stage PSTs are encouraged to go beyond just listing their procedural steps when solving a problem but explaining why those steps are taken and why and how they work. This activates the interplay between the conceptual and procedural understanding. Finally, PSTs are asked to solve a problem multiple ways and each unique approach is welcomed and valued. This gave them freedom to develop novel approaches and look beyond prescribed ways of doing mathematics. For example, many PSTs enter the courses knowing that the "invert and multiply" algorithm for dividing fractions works, but not why. This is a common problem, and we address it. But we also look for alternate solution and computation paths that allow PSTs to feel that they *can* compute the quotient using this algorithm rather than *must*. For the quotient $\frac{\frac{2}{5}}{\frac{3}{8}}$, we would want the PSTs to understand that they *can* multiply the numerator and denominator of this complex fraction by $\frac{3}{8}$, but we would also want them to appreciate that they *could* also multiply the numerator and denominator by 15. Being able to think beyond an algorithm or why an algorithm works supports a higher level of both conceptual and procedural understanding.

Design Background for the APEX Cycle: The APEX Cycle is a model of what Bass (2015) identifies as authentic mathematical work. He notes

Figure 9.1

The APEX Cycle

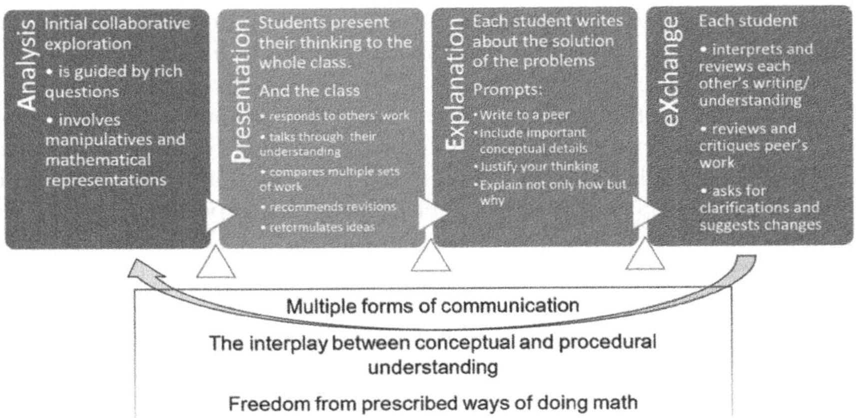

that "[m]athematical work generally progresses through a trajectory" that [Bass] would describe as: "Exploration—discovery—conjecture—proof—certification" (p. 631). Hiebert and his colleagues (1997) describe learning mathematics with understanding as a complex and dynamic phenomenon where we make connections to other things that we know. Hiebert et al. highlight the importance of reflection and communication cycle where "the process of reflection is central for cognitive psychology, and the process of communication is central for social cognition" (p. 5). Reflection would be in the form of metacognition where one thinks about what she or he is doing and why. Through communication, one's thinking would become somewhat visible to others. Thus, collective thinking would become possible which would encourage each member to reflect deeply about his or her own thinking.

Since we work with elementary preservice teachers, we also take into account Ball et al.'s (2008) complex notion of "Mathematical Knowledge for Teaching" (MKT). Ball and her colleagues argued that teaching a specific subject requires much more than knowing the content alone. When they described MKT, they partitioned it as subject matter knowledge and pedagogical content knowledge. Even the Subject Matter Knowledge (Common Content Knowledge, Horizon Content Knowledge, and Specialized Content Knowledge) and Pedagogical Content Knowledge (Knowledge of Content and Teaching, Knowledge of Content and Students, and Knowledge of Content and Curriculum) are unpacked. We argue that the APEX cycle allowed PSTs to reflect on this complex and dynamic knowledge for teaching and MKT by experiencing high-leverage practices and communicating and thinking about mathematics deeply.

Our APEX cycle integrates student-centered classroom environments focused on the use of lessons that employ Bass' exploration—discovery—conjecture—proof—certification and Hiebert et al.'s (1997) reflection-communication cycles. These cycles supported specific APEX strategies for PSTs to make sense of problems and persevere while exploring alternative solution paths as they worked on problems individually and in groups. For example, the critical analysis of the tasks and of each other's work, the presentation and defense of mathematical ideas, and the precise expression and representation of foundational content for themselves and others became regular practices of the course. The use of multiple representations and manipulatives helped PSTs to model and to use tools and representations strategically to help their thinking and communicating their ideas. Furthermore, the journal assignments engaged PSTs in meaningful mathematics and asked them to write detailed reflections of their understanding on the mathematics activities. This supported PSTs to construct viable arguments and attend to precision in their work and explanation. Finally, the exchange of journals across institutions allowed PSTs to critique the work and reasoning of other PSTs for the purpose of developing skills to review others' work in a positive, professional manner.

IMPLEMENTING THE APEX CYCLE

In this section the authors provide a description of how the first three parts of the APEX cycle were implemented differently in both institutions due to different scheduling of the courses. Then we will discuss the last stage of the APEX cycle collectively.

Wayne State University Case: The WSU courses met for four 50-minute periods each week. Two of these four periods were dedicated to the Analysis and Presentation components of the APEX Cycle. During the first of these dedicated periods, the PSTs collaborated on rich problem sets that engaged them in deep exploration of the conceptual underpinnings of elementary school mathematics (see Figure 9.2 for a sample problem set). For instance, PSTs explored mathematics important for their work as teachers such as whole number operations, and the conceptualizations, and representations of integers and rational numbers. During this work, we asked them to justify statements using definitions, properties, or theorems. The PSTs were asked to explain their thinking and to listen actively to the ideas of others. This helped the PSTs begin to appreciate the variety of solution paths available and to evaluate the affordances of each. One PST shared, "Working in small groups on the worksheet problems really helped me to understand the problems and to bounce ideas back and forth about different answers or how to get an answer *in various ways*" (emphasis added).

In the second period students presented their emerging understandings to their colleagues in a full-class setting. This public presentation addresses the call of the Conference Board of the Mathematical Sciences (2012) to have future teachers experience mathematics as a dynamic field where they have opportunities " to construct viable arguments, to listen carefully to other people's reasoning, and to discuss and critique it" (p. 33), so that they will be able to facilitate experiences for their future students. Afterwards, the students wrote detailed solutions to the problems where they focused on creating clear written explanations and mathematical representations in which they discussed both procedural and conceptual aspects of the work. Additionally, we asked PSTs to develop and make connections between several representations of a concept or procedure. This activity shown in Figure 9.2 invites PSTs to examine multidigit multiplication from several approaches and reflect on the connections among those approaches. Figure 9.3 shows one preservice teacher's work from her journal and how she connected the distributive law and the area model.

The color coding (not visible in Figure 9.3) and the explanations made her understanding visible to others. At the same time, it gave her a chance to reflect on her thinking in depth. She stated that

> there are many ways to solve the same problem. Each one provokes a different type of understanding and explains certain elements of the calculation. Also, it demonstrates how compacted our knowledge of the standard algorithm [of multiplication] really is, and we must unpack that knowledge in order to teach it to others.

Components of the problem set require PSTs to draw on specific domains of MKT (Ball et al., 2008). For example, with question four—inspired by the work of Ma (1999)—one can appreciate the Specialized Content Knowledge needed to diagnose the student error and formulate a response that targets the misconception (Ball et al., 2008). To address the student's misconception, one PST decided to make visible the reasons for "adding a zero on the second line" by connecting the standard algorithm to more expanded algorithms. Figure 9.4 shows a snippet of her journal write-up. This work is at the heart of learning mathematics with understanding (Hiebert et al., 1997).

The Henry Ford College Case: The HFC courses met for two 110-minute periods each week. Journal work extended over the last hour of one class and was continued for the first hour of the second class. Like the work at WSU, the first hour was devoted to preservice teachers' exploration of the problems (Analysis Stage). They worked individually and in small

Figure 9.2

Multiplication Worksheet

1.

a. Expand completely, using the distributive law:

$24 \times 58 = (20 + 4) \times (50 + 8)$

= ______________________

b. Label the rectangle below to illustrate this calculation.

2. Compute the product 24 × 58 using the expanded standard algorithm.

3. Compute the product 24 × 58 using the standard algorithm.

4. Your student has carried out the calculation using the standard algorithm as shown below. Explain what is wrong in their calculation. Choose one of the alternate algorithms (from numbers 1 and 2) to help address his misconception.

$$\begin{array}{r} 58 \\ \times 24 \\ \hline 232 \\ 116 \\ \hline 348 \end{array}$$

Figure 9.3

Connections Between the Distributive Law and the Area Model

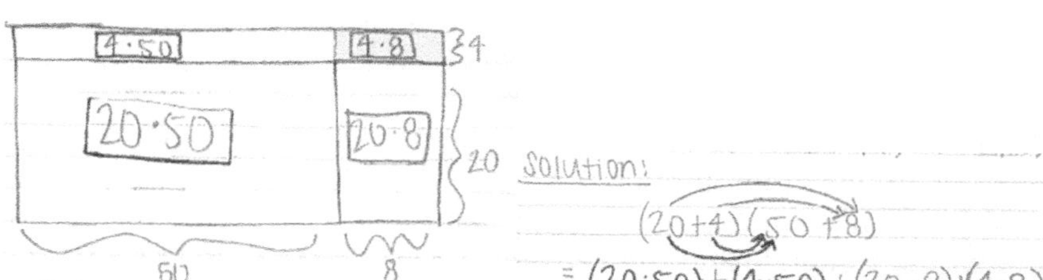

Figure 9.4

Connections Between the Standard and Expanded Multiplication Algorithms

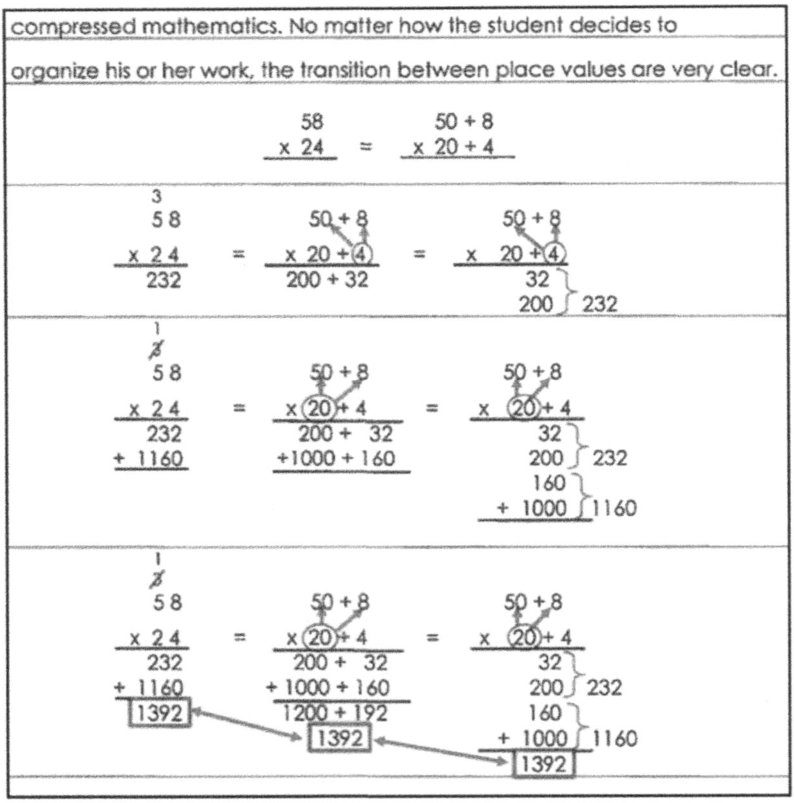

groups. Using the Whole Number Multiplication Worksheet, see Figure 9.2, PSTs worked through problem 1(a) and completed the multiplication of 24 x 58 using the distributive property. Samples of PSTs' work are found in Figures 9.5 and 9.6. These samples demonstrate varying levels of detail PSTs use during their initial drafts. PSTs compared their peers' work and determined that the more detailed work unpacked the mathematics making each partial product more visible.

For problem 1b, PSTs modeled the product using the given rectangle breaking out the multiplication of 24 x 58. They worked together to identify the expanded forms of 24 and place the 4 + 20 along the vertical edge of the rectangle and the expanded form of 58 as 50 + 8 along the horizontal edge. They identified the internal rectangles as the products of 4 x 50, 4 x 8, 20 x 50 and 20 x 8 (similar to the work that can be seen Figure 9.3). During this exploration phase, PSTs examined their work and found connections between the expanded standard algorithm resulting from the distributive property and the area model. The individual rectangles in Figure 9.3 from the area model were visible now in Figure 9.6 in the expanded standard algorithm. This type of fine-grained analysis of the problems continued. PSTs were prompted to draft their journals between the first- and second-class sessions so that they would be prepared to present their work to the class. When doing this initial draft, PSTs were prompted to think carefully about the mathematics and write an explanation as though they were working with a peer who had never engaged in this work. They were reassured that this was a draft and that they would have the opportunity to revise their work after class discussion the following class period. During the first hour of the second-class session, students presented their work to the class. PSTs displayed their work so that results could be analyzed for detail in language and representation.

Figure 9.5

One PST's Compressed Work

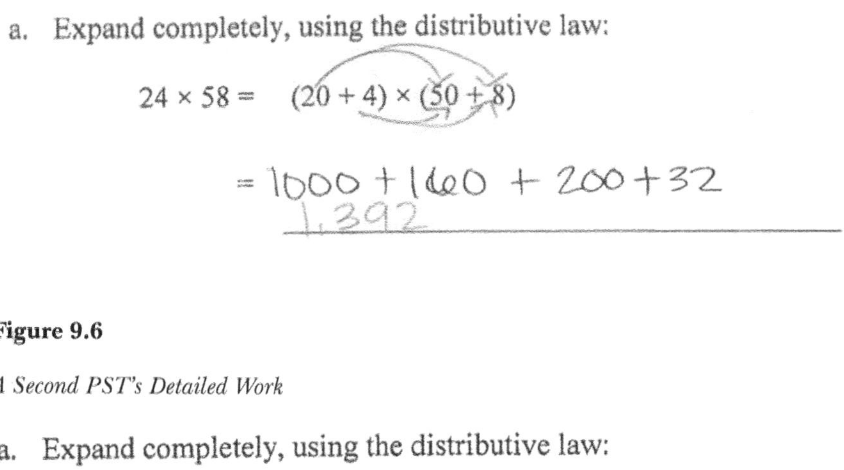

a. Expand completely, using the distributive law:

$24 \times 58 = (20 + 4) \times (50 + 8)$

$= 1000 + 160 + 200 + 32$

1,392

Figure 9.6

A Second PST's Detailed Work

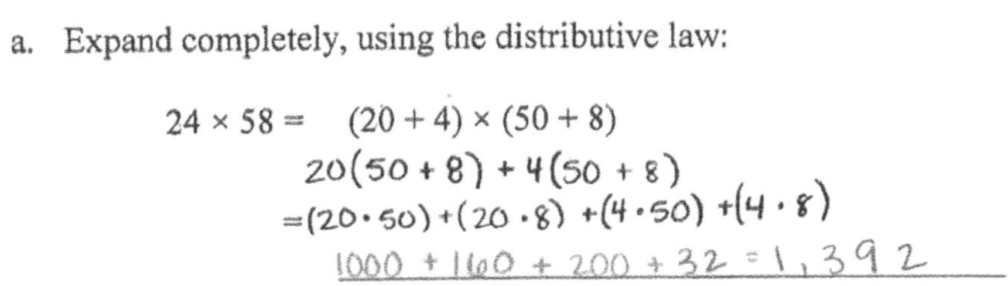

a. Expand completely, using the distributive law:

$24 \times 58 = (20 + 4) \times (50 + 8)$

$20(50 + 8) + 4(50 + 8)$

$= (20 \cdot 50) + (20 \cdot 8) + (4 \cdot 50) + (4 \cdot 8)$

$1000 + 160 + 200 + 32 = 1,392$

Conversation about the level of detail needed to make the multiplication of two-digit numbers understandable to students led to PST revision of their work. PST conversation continued with the level of detail needed for the area model in problem 1b and how the expanded and area models supported each other. Again, PSTs shared work that demonstrated varying levels of detail. Conversation focused on details that added to students' understanding of two-digit multiplication, connections between multiplication using the distributive property, area model and expanded standard algorithm in problem 1b (see Figures 9.7, 9.8 and 9.9 for samples of PST work). We see in these samples attention to the expansion of 24 and 58 in two ways. In Figure 9.7, the PST expands 24 to 10 + 10 + 4 and 58 to 10 + 10 + 10 + 10 + 10 + 8. She continues to multiply each term for 24 and each term for 58 resulting in four rectangles with area 40, one rectangle with area 32, 10 rectangles with area 100 and two rectangles with area 80. The PST who created Figure 8, expanded 24 to 20 + 4 and 58 to 50 + 8. Her representation had one rectangle representing the product of 4 x 50, one rectangle representing the product of 4 x 8, one rectangle representing the product of 20 x 50, which was partitioned to demonstrate ten groups of 100, and one rectangle representing the product of 20 x 8. These rectangles were labeled with the partial products and color coded. In Figure 9.9, the standard algorithm is unpacked so the PST can make the connections among different representations and algorithms. The PSTs drew representations providing different levels of detail which provides opportunities for discussion to answer the questions, "What mathematics do these representations unpack? How do these representations support the understanding of two-digit multiplication?" Like the work of the first day, PSTs reviewed the presented work, discussed the problems and made connections among the many procedures in problems one through three. When PSTs addressed problem 4, see Figure 9.10, the distinction between addressing procedures and concepts arose. The question asks PSTs to explain an error made by a student in the calculation of 24 x 58. Some PSTs addressed the error as the student not including a placeholder. Others noted that in the first set of products, the student was multiplying 4 x 8 and 4 x 50. However, in the second set of products, the student was multiplying 20 x 8 and 20 x 50. The multiplication by 20 required the need for a placeholder. This attention to multiplication by 20, or two sets of 10, raised the importance of conceptual understanding to the entire class. The PSTs worked carefully with language, representations, and concepts as they developed their answers for this task. As with the WSU classes, PSTs developed their understanding and language around the mathematical ideas on which the activity was focused. Our teaching of these classes was focused on achieving four goals: to model high-leverage teaching practices, to develop fundamental mathematical

Figure 9.7

One PSTs Detailed Representation of 24 x 58

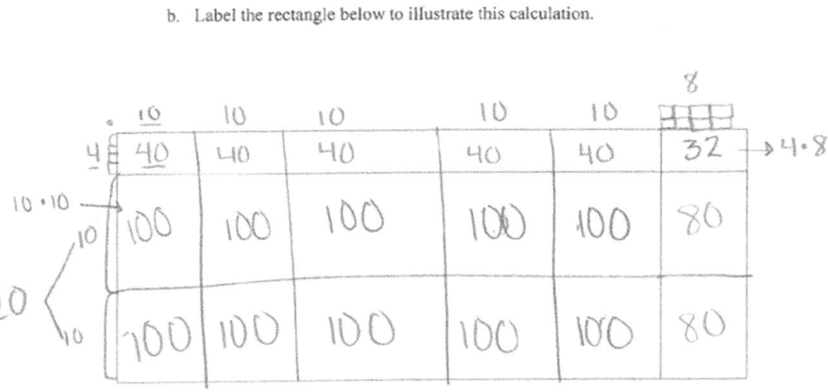

Figure 9.8

A Second PST's Detailed Representation of the Area Model

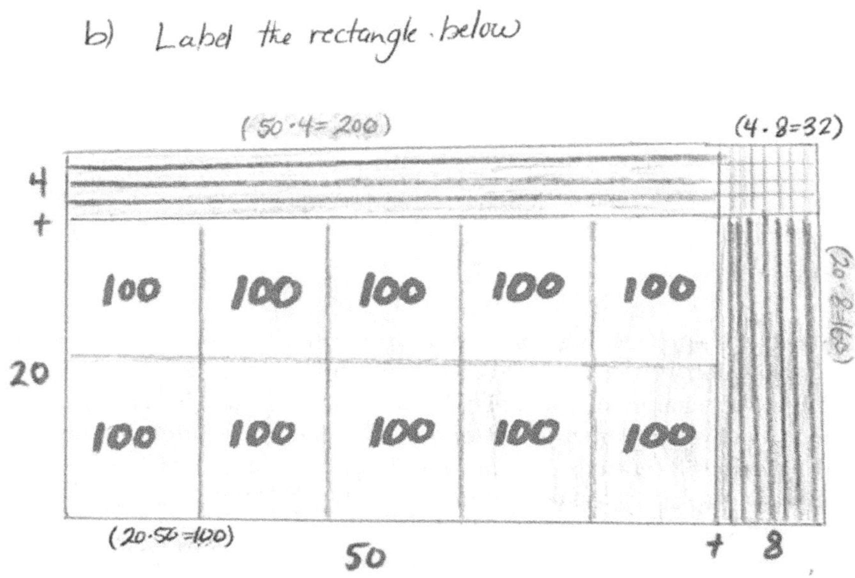

Figure 9.9

A PST's Work to Multiply Using the Expanded Standard Algorithm

2. Compute the product 24 × 58 using the expanded standard algorithm.

```
   24
 × 58
 ----
   32
  160
  200
 1000
 ----
1,392
```

① multiply 4 × 8 to get 32
② multiply 8 × 20 to get 160
③ multiply 50 × 4 to get 200
④ multiply 50 × 20 to get 1000
⑤ add all the products together to get 1,392.

*By making the 2 into 20 and the 5 into 50 helps us keep the place value of the numbers.

Figure 9.10

Problem 4

4. Your student has carried out the calculation using the standard algorithm as shown below. Explain what is wrong in their calculation. Choose one of the alternate algorithms (from numbers 1 and 2) to help address his misconception.

```
  58
×24
----
 232
 116
----
 348
```

content used for teaching children, to foster mathematical ways of work, and to develop written and oral communication skills in PSTs.

Digital Exchange of Journals: The courses were taught more than 25 times collectively at both institutions over five years; and students completed approximately 8 journals in each course—more than 200 journals overall. The main goal of asking preservice elementary teachers to write journals was to make visible their individual thought processes and to reflect on their own understanding. Two of these journals were exchanged in each term between institutions, so that PSTs could make sense of their peers' mathematics, critique work of other PSTs for the purpose of developing skills to review their peers' work in a positive and professional manner, and to gain a new perspective when reflecting on their own work. During the first two years, paper journals were exchanged and beginning in the third year we exchanged the journals digitally. Last academic year, we started with an in-class peer review followed by an interinstitutional exchange. Our purpose for having PSTs exchange between two institutions is to promote cross-institutional connections among the PSTs and to have them read the work of other PSTs who might be using different representations or language. Even though both groups of students were studying the same mathematics and working on the same assignments, perspectives and explanations may be different and may require more careful review.

Because of the timing of our course schedules, it took a few weeks to exchange journals between institutions, to have preservice teachers write responses, and to return the responses. We asked students to comment using the following guiding questions:

- Note the mathematical ideas that are being developed and discuss how reading about your peer's mathematical ideas helped your own understanding. In particular, how does the author make the work with place value visible in his/her writing?
- Compare/contrast how the journal author chose to address the elementary student's misconception with how you chose to address it. How does the explanation make visible place value in multiplication?
- Feel free to ask questions about the work.
- Lastly, as in the first exchange, if you would like to offer any additional encouragement or friendly words to the author, the end of the journal is an excellent place to do so.

PSTs stated that *writing* journals enhanced their mathematical thinking and they also thought differently about their writing. This may entail thinking about the mathematics in a different way or rethinking the explanation, recognizing that it needs more detail. One PST shared that "the journals were also a great help, as I've become more comfortable with my ability to explain varying mathematical concepts in detail." We saw that the *exchange* of the journals enabled PSTs to *deepen* their mathematical knowledge for teaching in multiple ways:

1. explain and communicate mathematics (mathematical thinking) in empowering ways (e.g., with effective use of representations),
2. acknowledging various ways of approaching a problem, and
3. making sense of someone else's mathematics.

In the interaction shown in Figure 9.11, dialog between two PSTs' highlights multiple portions of mathematical knowledge for teaching: comparison of various algorithms for multiplication, accepting and valuing multiples approaches to solve a problem, how strategically to use different methods to help students' thinking, knowing students' misconceptions and so on.

Figure 9.11

Digital Journal Exchange

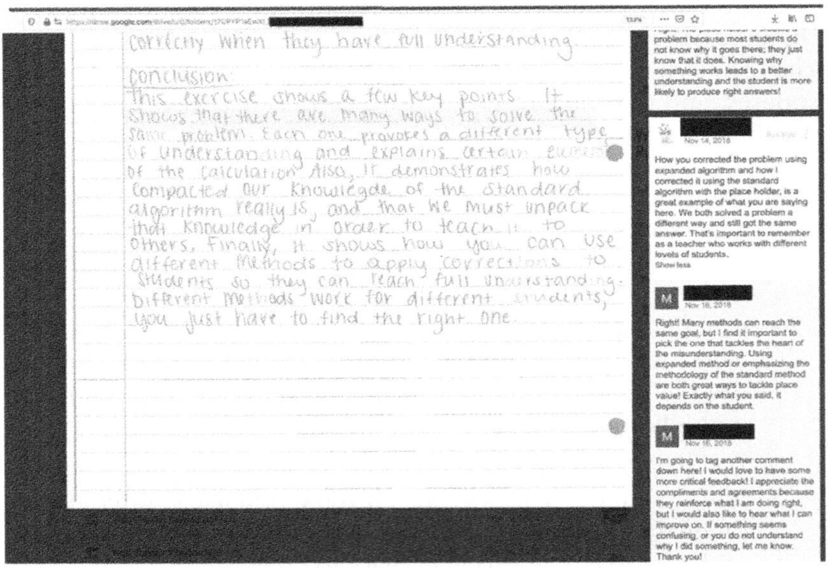

A PST reported that she can now "explain math in a way that not only makes sense to me but in a way that I will hopefully explain to future students." PSTs appreciated the process of making sense of someone else's mathematics: "I learned different ways that others solved the problems. It was unique to see different people's work and explanations." Hence, the whole experience furthered their mathematical knowledge for teaching and their openness to working with others. One student shared,

> Having someone else evaluate my work and provide suggestions for improvement allows me to see someone else's way of thinking, and may be a way to get new ideas I have never thought of before . . . In the future, going beyond comparing what I wrote to my peers, I should think deeper about the content and maybe look at important concepts that maybe even

> I missed in my own journal to help my peer and provide a perspective they might not have seen.

As we can see from this statement, the journal exchange made this PST think deeply about the content.

The digital platform changed the nature of the journal exchange. One-time commenting in the paper exchange still occurred in the digital exchange, but the digital platform opened the doors for peers across institutions to have a mathematical dialogue and made the exchange more efficient:

> At first I was hesitant with Google docs, but I am much more comfortable with operating that now. Knowing how to respectfully make comments on the work and be face to face with the reviewee can be valuable. Technology can be super-efficient, time saving and neater as well.

Our journal exchange activity produced valuable and unique mathematical learning experiences as stated in this following unsolicited feedback:

> I'm messaging you to let you know I actually really enjoyed doing the … exchange. I loved how we were able to send our sheets to others and reply to them, with them replying back in a matter of minutes/hours. It was a great experience. My friend from [the other school] was very nice and her feedback was great."

Finally, the will to improve one's mathematical communication, hence the mathematical reasoning, surfaced during these journal exchanges. PSTs asked each other to provide constructive feedback so they can improve their mathematical communication and thinking (see the final comment in Figure 9.11).

WHAT DID WE LEARN?

Our teaching shifted from a lecture/activity format to a more focused activity-based design, integrating several "high-leverage practices"—practices that are "crucial for pupils' learning and that beginning teachers can learn to perform proficiently" (Boerst et al., 2011, p. 2849). Specifically, we strategically used questioning and revoicing throughout our work to guide preservice teachers' exploration of problems that will elicit deep understanding of fundamental mathematics used for teaching children. Questioning was used for five specific purposes: (1) eliciting student thinking, (2) probing student answers, (3) directing students to listen and respond to others' ideas, (4) guiding students to make connections between multiple ideas, and (5) guiding students to reason mathematically (Boerst et al., 2011).

Our use of these pedagogical strategies modeled and provided an environment for PSTs to experience high-leverage teaching practices. These experiences promoted the development of their mathematical knowledge for teaching—knowledge of concepts, representations, and ways of talking about mathematics; making connections among multiple concepts, representations, and vocabulary; developing reasoning skills - looking for patterns, formulating conjectures, stating and using definitions, generalizing, justifying, proving—doing these and knowing that they are using these mathematical ways of work (Boerst et al., 2011). As preservice teachers engage in this challenging work, we found that they begin to identify the distinction between learning mathematics for themselves and learning mathematics for teaching (Ball et al., 2008). One student shared that "the journals were also a great help, as I've become more comfortable with my ability to explain varying mathematical concepts in detail." A second student noted that "instead of my brain working like a computer/calculator, I actually had to think and provide explanations." And a third student shared that journal "assignments improved my critical thinking skills, specifically in application of my knowledge." We found that PSTs found value in the journals and communicated the ways the journal writing, and critique helped them develop their knowledge of mathematics.

To our surprise, preservice teachers not only reviewed another's work, but also responded to reviews of their work. In many cases, conversations took place about the work via google docs. In one pairing, preservice teachers even spoke with each other on the phone to discuss their work on a journal prompt involving Base-5 addition. As an example of this type of exchange, consider an exchange which includes important questioning by the WSU preservice teacher about the HFC preservice teacher's approach to a problem (see Figure 9.12). The WSU preservice teacher asks, "How did you get your numbers?" The HFC preservice teacher responds,

> Good question. I used 5 + 2 instead of 7 because I need to find a pattern. From prior experience, I know that just adding 7 will not help me find a pattern. The main thing to creating a table is keeping the idea of a pattern in mind.

Later in the exchange, the preservice teachers acknowledge the help each has given and received (see last set of comments in Figure 9.12).

In another exchange, we see a WSU and HFC pair collaborating in a way that improved not only their mathematical work, but perspectives on how they will learn and grow as teachers. In Figure 9.13, a WSU PST comments on her HFC partner's detail, noting that she would like to use her partner's explanation. The HFC PST acknowledges the compliment and notes, "I am glad you want to use it. We as teachers learn from each other as well." Later in this journal exchange, the WSU student notes how much the HFC PST has developed in her writing of mathematics. Both PSTs communicate their excitement for learning and eagerness to learn more. This byproduct of the journal exchange is noteworthy.

CONCLUDING THOUGHTS

> *"I liked seeing my journal through new eyes. The process of reading and contemplating peer reviews helped me to think about the strengths and weaknesses of my journal writing procedure. Certain aspects of my journal that seemed so clear to me could be vague to someone else. This will inspire me to pay more attention to the language I use to describe things as well as the images I chose to illustrate my words."*

We believe that this quote summarizes the aim of the APEX cycle, the journal writing, and the journal exchange. The APEX cycle promoted deep thinking about the mathematics used for teaching children mathematics. Journal exchanges sparked meaningful conversations about mathematics

Figure 9.12

Mathematical Discussion During Digital Journal Exchange (Note: The Yellow Squares Show Highlighted Portions of the Journals on Which a Peer Commented)

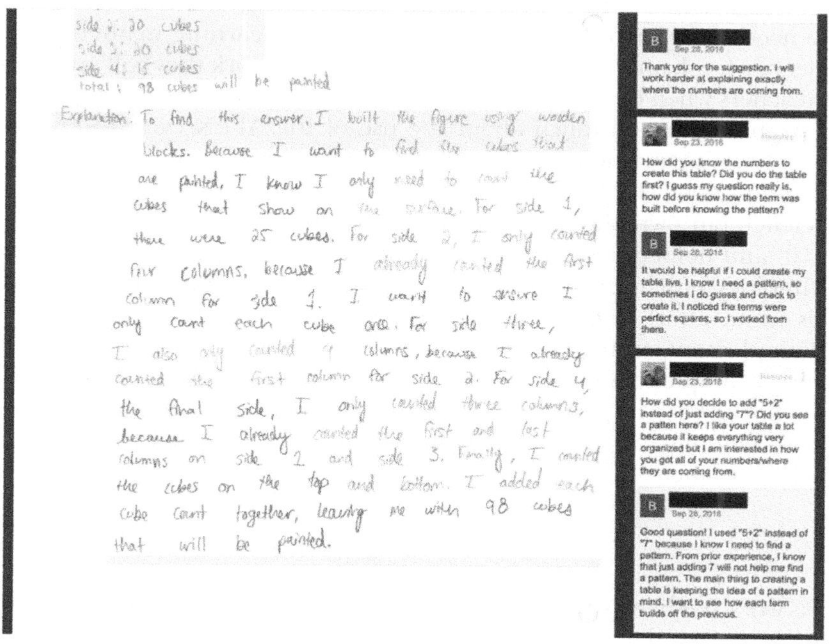

Figure 9.13

A Discussion During Digital Journal Exchange

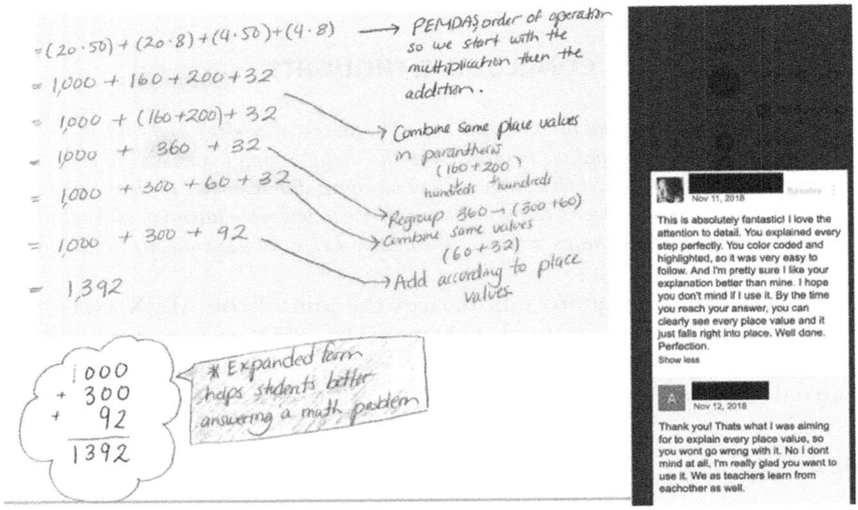

for teaching, catalyzed collegial conversations, made visible the importance of thinking about and communicating mathematics in several ways to meet the needs of learners, generated excitement for learning mathematics at a deeper level, and raised PSTs' awareness of the work they will engage in as teachers when developing lessons and sharing work with other teachers.

We have learned much about the use of the APEX cycle to enhance PSTs' learning mathematics for teaching. Our development of tasks, conversations about PSTs' work, and our work to deploy the APEX cycle have stretched our own Mathematical Knowledge for Teaching Teachers (Zopf, 2010) and helped make the cycle more effective. The implementation of technology enhanced the exchange, and we will explore other technologies in an effort to further develop PSTs' experiences. In addition, we will implement strategies as we launch new semesters providing PSTs with strong examples of journal writing and exchanges to help them learn from the experiences of past PSTs. We anticipate our work continuing to develop, helping PSTs learn mathematics in a way that will support their work as teachers.

ACKNOWLEDGMENTS

This work has been supported by two National Science Foundation grants: Division of Undergraduate Education, *Student Success Through Evidence-*

Based Pedagogies (SSTEPs) (DUE-1524878) and Robert Noyce Teacher Scholarship Program, *Noyce Mathematics Fellows, TeachDETROIT* (DUE-1540819).

REFERENCES

Bain, K. (2004). *What the best college teachers do.* Harvard Press. Quoted in http://blogs.ams.org/matheducation/2015/10/01/active-learning-in-mathematics-part-iii-teaching-techniques-and-environments/#sthash.CB93ed25.dpbs

Ball, D. L., Thames, M. H., & Phelps, G. (2008). Content knowledge for teaching: What makes it special? *Journal of Teacher Education, 59*(5), 389–407. https://doi.org/10.1177/0022487108324554

Bass, H. (2015). Mathematics and teaching. *Notices of the AMS, 62*(6), 630–636. https://doi.org/10.1090/noti1256

Boerst, T., Sleep, L., Ball, D., & Bass, H. (2011). Preparing teachers to lead mathematics discussions. *Teachers College Record, 113*(12), 2844–2877.

Hiebert, J. et al. (1997). *Making sense: Teaching and learning mathematics with understanding.* Heinemann.

Conference Board of the Mathematical Sciences. (2012). *The mathematical education of teachers II.* American Mathematical Society.

Ma, L. (1999). *Knowing and teaching elementary mathematics: Teachers' understanding of fundamental mathematics in China and the United States.* Lawrence Erlbaum Associates.

Mathematical Association of America. (n.d.). *A joint position statement of the Mathematical Association of America and the National Council of Teachers of Mathematics on teaching Calculus.* https://www.maa.org/programs/faculty-and-departments/curriculum-department-guidelines-recommendations/joint-statement-teaching-calculus.

Moses, R. P., & Cobb, C. E. (2001). *Radical equations.* Beacon.

Stigler, J. W., & Hiebert, J. (2004). Improving mathematics teaching. *Educational Leadership, 61*(5), 12–17.

Zopf, D. A. (2010). *Mathematical knowledge for teaching teachers: The mathematical work and knowledge entailed by teacher education* [Unpublished doctoral dissertation, University of Michigan]. Ann Arbor, MI.

CHAPTER 10

NUMBER TALKS

Engaging Students in Mathematical Reasoning

Dawn Marie Woods

INTRODUCTION

It is a typical day at Vista Elementary School.[1] The third-grade students in Mrs. Williams' class are coming in from recess and gathering in front of the whiteboard ready for a number talk. As Mrs. Williams displays the problem 43 – 8, students begin to solve the problem silently in their heads. Once students share their answers, the following interaction begins.[2]

1. Mrs. Williams: Raise your hand if you'd like to share your strategy. (Pause). Yes, Trent.
2. Trent: Um, what I did was I well, I took 40, I took 3 off of 43 and then I took 8 away from 40 and then I added, um, 3.
3. Mrs. Williams: Mm-hmm *(Mrs. Williams writes Trent's equations on the whiteboard as he explains his mathematical thinking.)*
4. Trent: Is 32, 35.
5. Mrs. Williams: Okay, so, okay, so you took the three off and you did 40 minus 8. And then what is that?
6. Trent: It is 32 ... 35.

Mathematics as the Science of Patterns:
Making the Invisible Visible to Students Through Teaching, pp. 205–228

7. Mrs. Williams: 32. *(Mrs. Williams writes 32 on the whiteboard.)*
8. Trent: And then I did 32 plus 3 equals 35. *(Mrs. Williams writes + 3 after the 32.)*
9. Mrs. Williams: Excellent. So why did you add this 3? *(Mrs. Williams points to the 3 in the second equation.)*
10. Trent: Because it's ... what, why did I add the 3?
11. Mrs. Williams: Yeah. So why did you add ... your answer's right.
12. Trent: Because I um, because I, I took it off of the 40, so I had to put it back on.
13. Mrs. Williams: Mm-hmm (affirmative) Love it. He took it out, but he had to put it back in to get his answer. A very good strategy.

This vignette is only a small portion of the number talk—an instructional activity where students mentally solve computation problems and then discuss how they made sense of the problem (e.g., Humphreys & Parker, 2015; Parrish, 2010/2014)—enacted by Mrs. Williams on this day. Still, it exemplifies how mathematical discussions support the development of accurate, efficient, and flexible strategies. Further, it highlights how number talks may support third-grade students to practice representing, justifying, and generalizing their mathematical thinking. For these reasons, number talks may provide sensemaking opportunities that build upon students' multiple mathematical knowledge bases (e.g., Turner et al., 2012) while bringing their mathematical thinking to new levels of understanding. Yet, research on mathematics teaching and learning (e.g., Hiebert et al., 2005) often shows missed opportunities for interactions, like illustrated in the vignette above, that make mathematical connections and relationships visible for students. This is a cause for concern since these interactions are where students are able to examine, critique, and validate their mathematical knowledge while building conceptual understanding (Wood, 1999).

In the present study, I investigate how number talks may be leveraged to support students to justify their mathematical thinking. Since this study is situated in a third-grade classroom, I view justification through the lens of a learning practice (Staples et al. 2012). Through this lens, justification becomes central to doing and learning mathematics. Hence, when teachers press students for *the why*, students justify or share their reasoning by explaining how they know that something is true (Stylianides, 2007). For example, in line 9 in the opening vignette, Mrs. Williams presses Trent to justify his reasoning when she asks, "So why did you add this 3?" Through the ensuing exchange, Trent explains, "Because I um, because I, I took it off of the 40, so I had to put it back on." In this moment, Trent not only justifies his mathematical reasoning, but also makes his compensation strategy transparent so that his classmates could also make sense of his solution strategy.

FRAMING IDEAS

In order to examine the role that number talks play in engaging students to justify their mathematical thinking, this section discusses the relevant literature revealing how students may become sensemakers of mathematics as they participate in number talks.

What Are Number Talks?

As revealed in the opening vignette, number talks are short mathematical discussions during which students solve problems and share how they made sense of the problem. Typically, number talks are designed to elicit students' mathematical reasoning by providing a routine where teachers (a) share a mathematical representation and provides students with enough time make sense of the task, (c) gather student responses, (d) facilitate a whole class discussion, and (e) summarize key ideas and conclusions (Humphreys & Parker, 2015; Parrish, 2010/2014). During number talks, students enthusiastically share their different strategies while being totally engaged and fascinated by the different solution strategies that emerge. Hence, number talks become a space where the teacher takes on the role of a learning coach supporting students in making sense of their own and their peers' mathematical reasoning. Because of this case for sensemaking, number talks are a popular instructional activity in many classrooms. Several books (e.g., Humphreys & Parker, 2015; Parker & Humphreys, 2018; Parrish, 2010/2014; Parrish & Dominick, 2016) and articles for practitioners (e.g., Berger, 2017; Parrish, 2011; Gerstenschlager & Strayer, 2019; Sun et al., 2018) suggest that number talks improve number sense, mental mathematics, and mathematical reasoning abilities. Yet, little rigorous research has be done to reveal the efficacy of number talks as an instructional practice (Matney et al., 2020).

The Role of Mental Computation During Number Talks

An integral part of number talks, as well as mathematics learning, is mental computation. Mental computation is characterized as producing an exact answer when a nontraditional algorithm or procedure is performed without using external devices such as paper and pencil (Reys, 1984; Sowder, 1988; Trafton, 1978). When students calculate mentally, they select from their repertoire of strategies and typically choose a process that is most efficient and effective for them (e.g., Blöte et al., 2000; Torbeyns & Verschaffel, 2016). The strategy that students select is often dependent upon the construction of the numerical problem (e.g., Siegler & Jenkins, 1989), word problem (e.g., Carpenter et al., 2015; Verschaffel & DeCorte, 1990), or everyday contextual problem (Nunes et al., 1993).

Although there are many benefits to mental computation as a visible part of an elementary mathematics program (Reys, 1984), particularly salient to number talks is that mental computation promotes greater understanding of the structure of numbers and their properties as students are encouraged to create ingenious ways of handling numbers (Cobb & Merkel, 1989; Reyes, 1984; Sowder, 1992). In fact, a growing body of research (e.g., Carpenter et al., 1998; Wood et al., 2006) shows that young students can develop their own efficient strategies spontaneously without instruction, suggesting that conceptual understanding of students is sometimes far ahead of classroom instruction (Blöte et al., 2000). Further, Kamii (1994) suggests that an early emphasis on learning algorithms is a *mathematical health hazard* that hinders children's numerical thinking by disrupting their development of number sense, as well as adding confusion to their sense of place value. Therefore, the role of mental computation during number talks may support students to understand how numbers work in a way that makes sense to them as they make decisions about how to solve a problem.

Creating Opportunities to Practice Mathematical Reasoning

Mathematical reasoning is an evolving process supporting students to conjecture, generalize, investigate, as well as develop and evaluate mathematical arguments (Lannin et al., 2011; Russell, 1999). In fact, it could be argued that mathematical reasoning is the essence of mathematical activity that not only lays the groundwork for mathematical proof, but could be utilized to build students' engagement in mathematics. To this end, the present study focuses on one aspect of mathematical reasoning: justification. In order to justify their mathematical reasoning, students need opportunities to make sense of mathematical ideas, properties, and established generalizations to show why an idea or solution is true, refute the validity of an idea, and/or give a mathematical defense of an idea that was challenged (Sowder & Harel, 1998; Stylianides, 2007). These opportunities to justify are not found in activities that apply rote procedures or memorized facts, but in activities where the conceptual underpinnings of mathematical reasoning are investigated. Particularly salient is that number talks may provide initial steps toward the use of justification processes since students have the ability to describe one's thought processes and then recognize whether or not the process makes sense when they build upon peers' examples, nonexamples, and thinking about the mathematical properties.

For example, recall in line 2 of the opening vignette when Trent had the opportunity to explain his thought process for solving the problem 43 – 8. As he explained his mathematical thinking, he was in the process of making sense of his compensation strategy as exemplified by the teacher's thoughtful prompting. Here, as Trent explained his process, he did not state that his claim was true until Mrs. Williams pressed and supported his thinking in line 11. At this point, Trent was able to then give a justification of why he needed to add the three in order to arrive at his final answer of 35. Hence, the number talk routine may have created a space for teachers and students to explore and develop the skills and process that comprise mathematical reasoning.

Conceptual Framework

The sociocultural theory of learning and the notion of communities of practice inform this study (Lave & Wenger, 1991; Wenger, 1998). Succinctly, people who collectively share the same passion or concern and engage in learning about their shared interest form a community of practice. Within this community, participants engage in joint activities and discussions as they build relationships that enable them to learn from each other (Wenger, 1998). Hence, learning is not seen as the acquisition of knowledge by individuals, but as an ongoing process of negotiation by members of the community. Particularly salient in regards to number talks, the practice of justification does not have an inherit form or meaning, but is given meaning within the context of the community (e.g., de Villiers, 1990; Staples et al., 2012). Therefore, approaching this work, I sought to understand justification as a learning practice in a third-grade classroom.

Since there has been little research on the efficacy of number talks as an instructional practice (Matney et al., 2020) and the research about justification as learning practice with K–12 classrooms is an emerging field (Staples et al., 2012), this study aims to build upon the contemporary literature by addressing the following research question: Can number talks be leveraged to support students to justify their mathematical thinking?

METHODS

The present study was part of a larger project investigating the development of ambitious mathematics instructional practices through the enactment of a daily number talk routine. During this project, teachers established math-talk learning communities as their roles shifted to supporting students in having meaningful conversations about mathematics (Woods, 2018). Further, exit tickets provided evidence that students' strategies shifted from a focus on sharing how they solved the problem to providing evidence of a mathematical relationship, and in some cases, justifying their mathematical reasoning (Woods, 2019). Particularly salient to the current study was that on these exit tickets Mrs. Williams' students tended to justify their mathematical thinking more than students in the other classes. Yet not every student told why an idea was true, signaling that there was still work to be done in developing ways to support students to develop conceptual understanding through justification processes. Therefore, this grounded theory study (Corbin & Strauss, 2007) aims to systematically develop a theory that may explain how justification becomes central to learning by investigating the factors, actions, and outcomes of students sharing their reasoning during number talks.

Participants

The participants in this study were Mrs. Williams and her third-grade students at Vista Elementary School. Vista Elementary School was situated within a predominately middle-class neighborhood within a large suburban district in the southwestern United States. This was Mrs. Williams first year to teach third-grade. Before teaching third grade, she spent three years teaching kindergarten. Also, number talks were not new to her. Before participating in this study, she attended two half-days of district training during previous school years, yet only enacted number talks once or twice per week. All of her students who returned parent permission forms were included in the study, resulting in nine female and seven male students in which 60% were White, 20% were Hispanic, 13% were Black, and 7% were two or more races.

Researcher Positionality

During this study, I was positioned as an elementary teacher since I was a former coteacher with one of the participants. This positioning provided me with the opportunity to work alongside these teachers as they enacted number talks in the classrooms. Also, this positioning provided me with the opportunity to investigate how number talks may have supported teachers to empower students to justify their mathematical thinking. For these reasons, I approached this study as a participant observer in order to provide an in-depth description of how students learned to justify their mathematical thinking during number talks (DeWalt & DeWalt, 2010; Jorgensen, 1989).

Data Collection and Analysis

The primary data were three video recordings and resulting transcripts of number talks enacted in Mrs. Williams' classroom during week 1, week 8, and week 15 of the 15-week study. In all, these three videos totaled approximately 47 minutes of recording time. All of the coding was conducted by the author and a trained graduate assistant. First, each of the three transcripts were open coded to determine the major categories of mathematical reasoning that emerged from the data (Creswell, 2013). During the analysis, mathematical reasoning—the development, justification, and use of mathematical generalizations (Russell, 1999)—emerged as a core phenomenon. Next, axial coding was developed to create categories around the core phenomenon (Strauss & Corbin, 1990). During this round of coding the procedure followed a process of (a) coding a subset of the transcripts and meeting to discuss the codes, (b) reconciliation of the codes, (c) double coding of the remaining transcripts, and (d) reconciliation of codes in order to reach intercoder agreement (Miles et al., 2014). Eventually, the analysis resulted in a visual picture (Morrow & Smith, 1995).

FINDINGS

As illustrated in the opening vignette, number talks opened up a learning space where Mrs. Williams supported her students to share their solution strategies. Here, a condition was created where she elicited her students' mathematical reasoning and pressed them to explain, elaborate, and clarify their thinking. As a consequence, students thinking was made visible as they revealed a deeper understanding of their reasoning and actions. In order to understand this phenomenon, the following section describes how Mrs. Williams supported her students to (a) tell why, (b) notice commonalities, and (c) move from justification to generalization during number talks. The section concludes by presenting a theoretical model engaging students in mathematical reasoning during number talks.

Pressing to Tell Why

The number talk enacted by Mrs. Williams during week 1 of the study was centered on the computation problem 47 + 53. As the number talk began, students had ample time to mentally solve the problem. They were also encouraged to come up with multiple solution strategies as they waited for all of their classmates to make sense of the task. It is after this quiet thinking time that the following interaction began.

1. Mrs. Williams: Noah, tell me, what is your strategy? How did you solve this?
2. Noah: So, put a 53 at the top. The 50 ... this is hard to explain.
3. Mrs. Williams: You want me to come back to you?
4. Noah: No, I know how to do it.... You put the 53 on top. It's like a t-chart.
5. Mrs. Williams: Yep, you're saying right, t-chart. Okay. So, tell me about that t-chart that you made. *(Mrs. Williams draws a t-chart on the whiteboard as Noah explains his mathematical thinking.)*
6. Noah: You put 53 on the top and the 47 on the bottom. 3 + 7 is 10.
7. Mrs. Williams: Okay. So, do I put the 10 down here? *(After Mrs. Williams stacks 53 on top of the 47 in the t-chart on the whiteboard, she points below the line as she asks where to write the 10.)*
8. Noah: No.
9. Mrs. Williams: Okay.
10. Noah: You take the one and you put it above the five.
11. Mrs. Williams: Okay. *(Mrs. Williams writes the one above the five.)*
12. Noah: And then 5 + 4 = 9. So, you put the 9 there and then it's 90.
13. Mrs. Williams: Oh, okay, let's look at some other strategies. *(Mrs. Williams writes the nine below the line.)*
14. Noah: Oh, wait. Oh wait, no, no, no. I forgot to add the 1.
15. Mrs. Williams: Oh, okay. So, then you did 5 + 4 and then you added?
16. Noah: One more. It equals ten.
17. Mrs. Williams: Okay, so....
18. Noah: And then you put the ... so it's 100.
19. Mrs. Williams: So, do I put both of those digits down here? *(Mrs. Williams writes the digits 10 below the line, the left of the 0 that was already there.)*
20. Noah: Yes.
21. Mrs. Williams: Why do I put both digits down here now? *(Mrs. Williams points below the line.)*
22. Noah: Because you can't ... there's nowhere else to add the 1.
23. Mrs. Williams: Very good. We've talked about that. That's when we put both digits down because there is no other place value to regroup it to.

In this vignette, it was not surprising that Noah was unsure about how to explain his solution strategy since this was one of his first experiences in describing his mathematical thinking in third grade. Yet, he persisted with his explanation since Mrs. Williams probed his thinking without hinting at a solution method. She then made mathematics visible in line 7 when she asked, "So, do I put the 10 down here?" as she recorded and represented

Noah's mathematical thinking on the whiteboard. This question highlighted a moment designed to extend Noah's thinking beyond the recall of a procedure since it pressed him to make sense of place value as he arrived at his answer. As the exchange continued, Mrs. Williams took the opportunity to emphasize the sensemaking process. In line 21 she asked, "Okay, why do I put both digits down here now?" This moment, or the press to tell "why," was an intervening condition that cued Noah to share his reasoning about where the numbers in his solution strategy came from. During this process of justification, Mrs. Williams asked an intentional question that supported Noah to justify his reasoning, thereby making mathematics visible to Noah and his classmates. Yet, during the number talk in week 1, this was the only moment where there was an opportunity for a student too tell why or reason about the answer to the computation problem.

Telling Why and Noticing Commonalities

Overtime, there was a shift in what students shared as they explained their strategies during the number talk routine. The number talk enacted by Mrs. Williams during week 8 was representative of this claim. On this day, the number talk centered on the computation problem, 8 x 25. As Mrs. Williams gathered students' responses, she heard that they all knew the answer was 200. At this point she asked, "Prove it: you say it's 200 how do you know?" It was here where the following exchange began.

1. Remi: So, I tried a different strategy, so I did 25 … you first have to put, you have to do what eight times five is.
2. Mrs. Williams: Okay, so you did eight times five. *(Mrs. Williams writes the equation* 8 x 5 *on the whiteboard.)*
3. Remi: Cause I know it equals 40.
4. Mrs. Williams: Okay. *(Mrs. Williams completes the equation* 8 x 5 = 40.*)*
5. Remi: I did eight times two.
6. Mrs. Williams: Okay. Eight times two. *(Mrs. Williams writes the equation* 8 x 2 *on the whiteboard.)*
7. Remi: That equals 16, so … *(Mrs. Williams completes the equation* 8 x 2 = 16.*)*
8. Mrs. Williams: Okay.
9. Remi: I'm confused.
10. Mrs. Williams: She says she's lost. Who can help her out? We've got eight times five equals 40, eight times two equals 16. Who can complete the next few steps?
11. Evan: So, she actually, when she multiplied the eight times two it was actually supposed to be eight times 20.

12. Mrs. Williams: Why is it eight times 20? *(Mrs. Williams corrects the expression* 8 x 2 *to read* 8 x 20 *and erases the 16.)*
13. Evan: Because that two is in the tens place.
14. Mrs. Williams: Because this two is in the tens place. So, eight times 20 equals ...
15. Evan: 160.
16. Mrs. Williams: 160 good. *(Mrs. Williams rewrites the equation to read* 8 x 20 = 160.*)*
17. Evan: We just can add those products. So, 160 + 40 and get her answer to 200.
18. Mrs. Williams: 200. Good. *(Mrs. Williams writes* 160 + 40 = *200 on the whiteboard.)* So, Remi, you see, so that eight times two.... You broke it up. You broke the 25 into twenty plus five just make sure you look at the place value.

During this discussion, Remi decided to share a strategy that was new to her, signaling that her thinking and resulting explanation were not solidified. She knew that eight times five was equal to 40 and two times eight was 16. At this point, Remi became confused because she knew that 40 plus 16 was not 200. Because of Remi's hesitation, Mrs. Williams chose to open up the conversation for other students to build upon the thinking that Remi began. It is here where Evan, in line 11, explained, "So, she actually, when she multiplied the eight times two it was actually supposed to be eight times 20." Then Mrs. Williams encouraged justification by asking, "Why is it eight times 20?" This move opened up the opportunity for reasoning about where the numbers came from in this solution strategy. Because of this moment of justification, all the students heard the explanation, "Because that two is in the tens place," which made the importance of place value visible for every student.

Mrs. Williams prompt for justification may have reminded students about the importance of making mathematics visible by sharing why an idea was true when they presented their mathematical thinking. This prompt may have also played a role in the following solution strategy. Here, Owen volunteered his mathematical thinking, as well as his reasoning, when Mrs. Williams prompted for a different solution strategy.

1. Owen: I got the eight and split them into two, fours.
2. Mrs. Williams: Okay. *(Mrs. Williams first writes the equation* 8 x 25 *on the whiteboard and then draws two lines from the eight and writes a four at the end of each one.)*
3. Owen: So, I know that four quarters equals 100. So, I thought four times 25 equals 100. So, I did the same thing again: four times 25 equals 100. Then I plussed those two hundreds to get 200.

4. Mrs. Williams: Who can repeat what Owen just said. He noticed something very important. *(Mrs. Williams writes x 25 next to each 4, and then completes the equation to read,* 4 x 25 = 100.*)*
5. Mrs. Williams: What did he notice, Joseph?
6. Joseph: That four quarters of 25 equals 100 dollars, my same exact strategy.

This exchange illustrated a shift in how students were sharing their mathematical thinking. In particular, Owen spontaneously shared his reasoning about where the numbers in his solution strategy came from he stated, "I know that four quarters equals 100." This moment signified that the process of justification, or telling why, was becoming central to learning since Owen automatically shared why his idea was true.

In the previous exchange, Mrs. Williams in line 4 highlighted a sensemaking strategy of using information that they knew was true—four quarters equal 100—or identifying a commonality that extended the reasoning beyond the range in which it originated. This commonality or generalization seemed to play a key role in Lily's solution strategy.

1. Lily: Since I knew that a quarter was 25, I actually I did that and I, I did 25 plus 25 and it equaled 50. And then I did another 25 and another 25 and it equaled 50. And another 25 and another 25 to make 50. And another 25 and 25 to make 50. And I added all those 50's up.
2. Mrs. Williams: *(Mrs. Williams writes* 25 + 25 = 50 *four times on the whiteboard.)* Okay.
3. Lily: And then I added two of the 50's and it equaled 100. Then I added the other two 50's and it equaled 100. And then I added 100 plus 100.
4. Mrs. Williams: Very good. Okay, again, she's using the 25 she's thinking of it as quarter. Very good because if you have two quarters you have 50 cents. Excellent. *(Mrs. Williams writes* 50 + 50 = 100 *two times on the whiteboard. Then she wrote* 100 + 100 = 200.*)*

During this solution strategy, Lily shared that she accepted the generalization that 25 was the same as a quarter. This implied that she just knew that two quarters, or 25 plus 25 was 50. She then applied this idea to her solution strategy, carefully accounting for all of the quarters in the problem. Similarly, Theo, who was next to volunteer his solution strategy, built his strategy on the shared generalization that 25 was the same as a quarter.

1. Theo: I kinda did the same thing that she did but not the same. I took the quarters and added them. So, 25 plus 25 plus 25 plus 25 plus 25 plus 25 plus 25 plus 25 and that equaled 200.
2. Mrs. Williams: So, you added all the quarters up together, or 25's. Okay, excellent.
3. *(Mrs. Williams writes* 25 + 25 + 25 + 25 +25 + 25 + 25 + 25 = 200 *on the whiteboard.)*

In line 1 of this exchange, Theo added up eight quarters, ultimately reaching the answer of 200. He not only used what he knew about quarters but applied it in way that was different from the solution strategies that were shared before his turn.

In sum, Mrs. Williams enacted the week 8 number talk in a way that supported her students to make sense of the problem by probing thinking, making mathematics visible, and also encouraging justification. When she slowed down the conversation and pressed Evan for the why, she not only made mathematics visible to the students in the classroom, but also encouraged Evan to reveal a deeper understanding of his reasoning and actions. Further, this moment singled to her students that the process of justification was important, leading Owen to justify his mathematical reasoning without a prompt. The result of this process, was a generalization about the number 25 being the similar to a quarter. This shared commonality continued to make mathematical sensemaking visible, thereby leading Lily and Theo to eagerly apply what they already knew to solve the problem in ways that may have validated their thinking.

From Justification to Generalization During Number Talks

The number talk enacted by Mrs. Williams in week 15, continued to support students in reasoning about where numbers came from. There were two parts to the number talk on this day. The first part centered on the computation problem 23 + 44 + 7 and was a student-led number talk facilitated by Ethan. This number talk began with Scarlett sharing her thinking and then continued with Penelope telling how she solved the problem. Next, Evan shared his solution strategy. In his turn, he defended his strategy.

1. Evan: What I did to make it more easy, instead of doing 23 plus 44, I did 23 plus seven because 23 plus seven is 30. So, I could just add it.
2. Ethan: 23 plus 7, okay. Equals. *(Ethan writes* 23 + 7 *on the whiteboard.)*

3. Evan: 30. Then, 30 plus 44 and that equals 74.
4. Ethan: 30 plus 44. And that equals what? *(Ethan completes the equation* 23 + 7 = 30 *and then writes the expression* 30 + 44 *on the whiteboard.)*
5. Evan: 74.
6. Ethan: Good job *(Ethan completes the equation by adding = 74 after the expression.)*
7. Mrs. Williams: Hey. Can I interrupt really quick, Ethan?
8. Ethan: Mm-hmm (affirmative).
9. Mrs. Williams: What did Evan do? What is that called when you make, when you put two numbers together and you get a number like this?
10. Noah: Sum.
11. Lily: Friendly numbers.
12. Mrs. Williams: Yes, friendly numbers, but it is also sum. A sum is two numbers added together and your answer to an addition problem. But it's also when a number may end in a zero, or we've even talked about fives, are easy for us to do because we can count by fives and zeroes pretty easily. So, Evan made a nice friendly number, and then took the 44 and added it to his 30.

In line 9, as a consequence to Evan's justification, Mrs. Williams took the opportunity to make mathematics visible by highlighting the idea of friendly numbers. In fact, this idea of friendly numbers began to go through the process of generalization as it was applied to the next solution strategy. In the following exchange, Bryson explained his solution strategy.

1. Bryson: So, I got a seven and a four.
2. Ethan: So, you got a seven and a four. So, 74?
3. Mrs. Williams: Seven plus four or 74?
4. Bryson: Seven plus four.
5. Bryson: And I got 11. *(Ethan writes the equation* 7 + 4 = 11 *on the whiteboard.)* So, then I got 23 plus 40.
6. Ethan: 23 plus 40.
7. Bryson: And I got 63.
8. Ethan: Equals 63. *(Ethan writes the* 23 + 40 = 63 *on the whiteboard.)*
9. Bryson: And I did 11 plus 63 equals 74.
10. Ethan: 11 plus 63 ... *(Ethan writes* 11 + 63 *on the whiteboard.)*
11. Jayla: Wait, where did you get 11?
12. Bryson: I got the seven and the–
13. Ethan: Wait, 74?
14. Mrs. Williams: 11 plus 63 equals–
15. Evan: He got the 11 from the seven and the four. He just ...

16. Mrs. Williams: Okay.
17. Student: He got the same answer.
18. Mrs. Williams: So, you got the same answer? So, let's talk about that. Where did the seven and this four come from Bryson? Where did this seven and this four come from to get eleven?
19. Bryson: It came, the, uh, one's place. The seven on that side and the four right there.
20. Mrs. Williams: The four in the one's place?
21. Bryson: Yeah. The, those, uh, those are both ones and they both equal 11.
22. Mrs. Williams: So then 44 is now left as a 40 and then you can do 40 plus 23, which would give you 63. And then if you add your 11 plus 63, you get 74. *(Ethan completes the equation* 11 + 63 = 74 *on the whiteboard.)* Okay. Excellent. Does that make a little bit more sense, Jayla?
23. Jayla: Mm-hmm (affirmative).

During this conversation, Bryson extended the reasoning of friendly numbers beyond the range in which it originated as he applied the reasoning to his own solution strategy. In line 1, Bryson decomposed 44 in in his mind into a 40 and a 4 since these numbers were easier to work with. Then he took the 4 and added it to the 7. This gave him the sum of 11. Then, in line 5, Bryson explained that he added the 40 to the 23. This gave him 63. Finally, in line 9, he described that he added the 11 to the 63 in order to arrive at the answer of 74. However, his explanation was not as clear as it could be, causing Jayla, in line 11, to ask, "Wait, where did you get 11?" This intervening condition ultimately led to Bryson's justification in line 19 which stated, "It came, the, uh, one's place. The seven on that side and the four right there." In this moment, Mrs. Williams facilitated the conversation in a way that made the mathematics visible. She summarized the result of the conversation in line 22 so that Jayla and her classmates had the opportunity to understand how Bryson applied the concept of friendly numbers, as well as decomposition, to his solution strategy.

After Bryson shared his solution strategy, Mrs. Williams shifted to facilitating the second part of the number talk. Here, she asked students to solve the computation problem, 123 – 25 – 7. This part of the number talk focused on students generalizing or extending the reasoning of friendly numbers as they extended it to include the strategy of compensation or reformulating the problem so that it can be computed mentally. Here, Owen began his solution strategy by extending the concept of friendly numbers as he rounded numbers to make a number easier to subtract.

1. Owen: I round 123 to 125.
2. Mrs. Williams: Okay. So, you rounded 123 to 125.
3. Owen: Then did 125 minus 25.
4. Mrs. Williams: Okay. *(Mrs. Williams writes* 125 – 25 *on the whiteboard.)*
5. Owen: Then I got 100. Since I had first turned 123, I minused two.
6. Mrs. Williams: Okay, so since you increased this number, since you rounded it up by two, then you subtracted the two.
7. Owen: Then I got 98.
8. Mrs. Williams: Okay, then you got 98. *(Mrs. Williams completes the equation,* 125 – 25 = 100 *and then writes* 100 – 2 = 98 *on the whiteboard.)*
9. Owen: Then I got seven minus four.
10. Mrs. Williams: Say that again.
11. Owen: Seven minus four.
12. Mrs. Williams: Seven minus four. *(Mrs. Williams writes* 7 – 4.*)*
13. Owen: No. Seven minus five.
14. Mrs. Williams: Seven minus five. *(Mrs. Williams corrects the expression to* 7 – 5.*)*
15. Owen: And I got two. And then I added 98 plus two. And I got one hundred.
16. Mrs. Williams: *(Mrs. Williams completes the equation* 7 – 5 = 2. *Then she writes* 98 + 2 = 100 *on the whiteboard.)* Okay.
17. Owen: And then since I subtracted that five, I did 100 plus five and that's your answer.
18. Mrs. Williams: So, you wanted to get to a two to make 98 plus two to 100. Oh, okay. Okay, I got you. So, you knew seven minus five would equal two. *(Mrs. Williams writes* 100 + 5 = 105 *on the whiteboard.)*
19. Owen: Then I split the seven into two numbers.
20. Mrs. Williams: Okay. You split the seven into a five and a two. Gotcha. Okay.

In this exchange, Owen generalized the concept of friendly numbers in order to make subtraction easier. In line 1 Owen rounded 123 to 125 so that it would be easier to subtract 25. Then in line 5 he explained that he had to subtract the two that he added in order to compensate for the two that he added in order to make the number easier to mentally subtract. In line 6, Mrs. Williams slowed down the conversation in order to make this compensation strategy visible to all students. Then, beginning in line 9, Owen decided to decompose the seven into a five and two in order to make the numbers friendly. Here, he added the two to 98 to get 100 and then added the five to reach the final sum of 105.

Next, Evan shared that he was going to use the friendly number strategy, like he did when he shared his solution in part one of this number talk.

1. Evan: What I did, I wanted to make it, like, when the last one earlier, to make a nice, friendly number.
2. Mrs. Williams: Mm-hmm (affirmative).
3. Evan: I did 123 plus seven.
4. Mrs. Williams: 123 plus seven. Which equals?
5. Evan: 130. So, I can make it nice and friendly.
6. Mrs. Williams: So, plus seven is 130. *(Mrs. Williams writes* 123 + 7 = 130 *on the whiteboard.)*
7. Evan: I did 130 minus 25. And you can't do zero minus five so I regrouped one 10 from the three and then I made it a two. Then I made the zero a ten and ten minus five is five. And then two minus two is two and then I got 105.
8. Mrs. Williams: Okay. *(Mrs. Williams writes* 130 – 25 = 105 *on the whiteboard.)*

In this discussion, Evan made the numbers friendly by applying the associative property of addition when he added 123 plus seven. Although this was a missed opportunity to discuss the property of addition that was used to make the friendly number, the result or sum was 130, which was "nice and friendly" as stated in line 5. Also, in line 9, Evan made his thinking visible as he revealed a deeper understanding of where numbers came from. In fact, he justified his thinking as he described the process of regrouping that he used in order to subtract 25 from 130, resulting in the difference of 105.

Then, Paisley created friendly numbers by compensating. Like Owen, she chose to round 123, but instead of rounding up to 125, she took off the three.

1. Paisley: Okay. Um, so. I just did, I just thought 123, but I took the three off. Then I did 120 minus 25 and then I got, I got, uh, 95.
2. Mrs. Williams: Okay. *(Mrs. Williams writes* 120 – 25 = 95 *on the whiteboard.)*
3. Paisley: And then I still had three. So, I got the three and I did 95 plus three. I got 98. And I still have the two so 98 plus two is 100 hundred. Wait, wait, wait. No. No two. No two.
4. Mrs. Williams: 98. *(Mrs. Williams writes* 95 + 3 = 98 *on the whiteboard.)*
5. Paisley: 98 plus seven and then I got 105.
6. Mrs. Williams: Okay. *(Mrs. Williams writes* 98 + 7 = 105 *on the whiteboard.)*

During Paisley's explanation of her solution strategy, she applied the idea of creating a friendly number through compensation. In this moment, she revealed a deeper understanding of where numbers came from when she stated "and then I still had three" in line 3. This process of justification then prompted her to share how she added the three on to 95 in order to arrive at the sum of 98. She then carefully thought through her solution strategy and realized that she still needed to add the 7 in order to reach the sum of 105.

The final solution strategy was shared by Calista. At the start, her strategy was similar to Paisley's in that she rounded 123 to 120 in order to make a friendly number.

1. Calista: So, I decided ... I took 123 and rounded it to 120. And then I added three plus seven equals 10.
2. Mrs. Williams: Um, oh. I thought you said, three plus seven, which equals 10. *(Mrs. Williams writes* 3 + 7 = 10 *on the whiteboard.)*
3. Calista: And then, um, I subtracted 120 and 25 and then, um, that–
4. Mrs. Williams: 120 minus what?
5. Calista: Um, 25.
6. Mrs. Williams: Minus 25.
7. Calista: It was 95.
8. Mrs. Williams: Okay, 95. *(Mrs. Williams writes* 120 – 25 = 95 *on the whiteboard.)*
9. Calista: And then I add 10 to 95 and it was 105 and that's my answer.
10. Mrs. Williams: Okay. *(Mrs. Williams writes* 95 + 10 = 105 *on the whiteboard.)*

As Calista explained her strategy, she shared how she handled the three that she took off of the 123 when she rounded. Here, Calista combined the three with the seven to make a quick ten. Next, she subtracted 25 from 120, resulting in the difference of 95. Then she added the quick 10 to 95, resulting in the sum of 105. During this process, Calista used compensation, combined with the creation of a quick 10, to make the numbers friendly, or easier to mentally calculate.

The number talk in week 15 was mathematically rich because students had the opportunity to deeply engage with mathematics. Particularly salient, was how students applied the mathematical concepts and strategies, such as friendly numbers, decomposition, and compensation that were made visible to their own solution strategies. Further, the process of generalization became a consequence of justification since students' reasoning provided the insight into how a mathematical solution was true.

A Theoretical Model for Leveraging Number Talks to Facilitate Mathematical Reasoning

Across weeks 1, 8, and 15 the learning space created by number talks became increasingly powerful as Mrs. Williams, and eventually her students, slowed down the conversation to make mathematics visible. The move to make mathematics visible encouraged her students to reveal a deeper understanding of their reasoning and actions. Figure 10.1 illustrates a theoretical model describing how number talks produced a causal condition supporting mathematical reasoning. When Mrs. Williams pressed her students to justify their mathematical thinking, she generated an intervening condition designed to slow down the conversation in a way that provided students the opportunity to validate their work. Further, as students justified their mathematical reasoning, a consequence was making mathematics visible to all students. In this space, opportunities to discuss connections among mathematical ideas became a norm. This discussion supported students to identify commonalities and supported them to begin the process of generalization as they extended the reasoning to their own solution strategies during the remainder of the number talk.

Figure 10.1

Leveraging Number Talks to Engage Students in Mathematical Reasoning

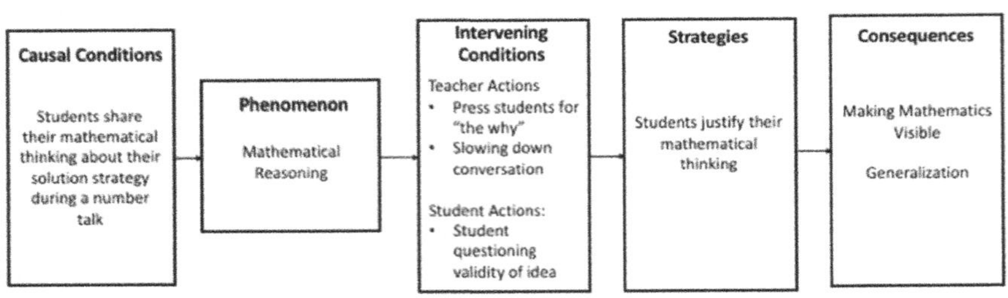

DISCUSSION

The findings from the analysis revealed that enacting number talks in Mrs. Williams' third-grade classroom established a learning space where mathematics was made visible to the community of learners. In this learning space, the teacher, or other community members, pressed for the why in a way that slowed down the conversation so that mathematical thinking could be justified. As a result of the discussion, members of the learning community began to extend reasoning, or generalized mathematical ideas by applying it to their own solution strategies. In this section, I elaborate on this overall finding, while connecting the results with prior literature.

The Role of Mental Computation Practicing Mathematical Reasoning

As described earlier, an integral part of number talks was mental computation. As students practiced mental computation on a regular basis, it afforded the opportunity to select a nontraditional algorithm or procedure that was most efficient and effective for them (e.g., Blöte et al., 2000; Reyes, 1984 Torbeyns & Verschaffel, 2016). In these data there was a noticeable shift in the procedures that students selected over the course of the study. Recall, in week 1, that Noah described how he used a t-chart, or the traditional algorithm, to solve the computation problem 47 + 53. Then, in week 8, Remi decomposed the 25 into 20 and five so that she could apply the distributive property of multiplication in order to multiply each number by eight to solve the computation problem of 8 x 25. Also, during this week, Owen built upon what he knew—"four quarters equals 100"—in order to solve the problem. Because of the highlighting of this commonality—or generalization about 25 as a quarter—students were encouraged to apply the reasoning to their own solution strategies. Last, in week 15, students generalized the concept of friendly numbers to their own solution strategies as they applied the associative property of addition, as well as decomposition and compensation strategies to mentally solve the computation problems of 23 + 44 + 7 and 123 – 25 + 7 In sum, the procedures that students selected moved away from the traditional algorithm to procedures that generated imaginative ways of handling numbers, which, in turn, uncovered the structure of numbers and their properties (Cobb & Merkel, 1989; Reyes, 1984; Sowder, 1992). This finding corroborated that students' conceptual understanding was often ahead of classroom instruction (Blöte et al., 2000). Further, this finding was similar to a growing body of research (e.g., Carpenter et al., 1998; Pinto & Cañadas, 2019; Wood et al., 2006) illustrating that students developed efficient strategies without direct instruction of traditional algorithms.

Justification as a Learning Practice

Consistent with the literature, when justification was viewed as a learning practice, students were pressed for the why in hope that they would express higher levels of mathematical reasoning (Staples et al., 2012). Particularly salient to this study was that opportunities to justify were found in the moments when the conceptual underpinnings of mathematical reasoning were investigated because of the role mental computation played in the invention of nontraditional solution strategies. Further, number talks may have provided initial steps toward the use of justification processes, and as a consequence generalization processes, because students described their thought processes and then recognized whether or not the process made sense while building upon shared examples, nonexamples, and thinking about the structure of numbers and mathematical properties.

Implications for Practice

This study provided evidence that the number talks supported students to select a nontraditional algorithm or procedure that was most efficient and effective for them as they mentally solved the computation problem. In turn, these solution strategies provided opportunities for community members to press for the why, and as a consequence, were able to extend the reasoning—or generalize—to other similar situations. Taken together, number talks supported Mrs. Williams in engaging her students to reason about where numbers came from in a way that supported them in making sense of mathematical ideas. Further, as number talks encouraged reflection and justification, equitable outcomes among diverse populations have been supported (Bieda & Staples, 2020; Boaler, 2006; Boalar & Staples, 2008).

Limitations & Directions for Future Research

As with most school-based research there are limitations to these findings. First, there may be experimenter effects because I regularly visited the classrooms to observe and video record number talks. I also worked with a subset of students. Also, a question remains how these findings might compare to other teachers, across different content and grade levels, as well as different school settings. Despite these limitations, this study confirms prior research on mental math and justification as a learning practice. Yet, future research is needed to better understand the efficacy of number talks as an equity-based instructional practice and the impact that they may have on student learning.

CONCLUSION

The findings from this study suggested that number talks provided students with opportunities to be sensemakers of mathematics revealing that number talks can be leveraged to support students to justify their mathematical thinking, thereby adding to the literature on the efficacy of number talks as an instructional practice. This study also revealed how justification, as learning practice, enhanced students' understanding and proficiency at doing mathematics (Staples et al., 2012).

Particularly salient, as students invented strategies and these solutions were valued and discussed within the classroom community, students had opportunity to make sense of mathematics in a way that was consistent with the research on ways that students learn mathematics with understanding (e.g., Bransford et al., 1999; Carpenter et al., 2015; Lester, 1994). Further, as students reasoned about where the numbers came from, a consequence was the creation of learning space where mathematics was made visible in a way that supported members of the learning community to apply the generalizations to their own solution strategies.

NOTES

1. The school and participants have been provided with pseudonyms.
2. This number talk enactment took place during week three of the study.

REFERENCES

Berger, A. M. (2017). Using number talks to build procedural fluency through conceptual understanding. *Ohio Journal of School Mathematics, 75*, 1–7.

Bieda, K. N., & Staples, M. (2020). Justification as an equity practice. *Mathematics Teacher: Learning and Teaching PK–12, 113*(2), 102–108. https://doi.org/10.5951/mtlt.2019.0148

Blöte, A. W., Klein, A. S., & Beishuizen, M. (2000). Mental computation and conceptual understanding. *Learning and Instruction, 10*(3), 221–247. https://doi.org/10.1016/S0959-4752(99)00028-6

Boaler, J. (2006). Opening our ideas: How a de-tracked math approach promoted respect, responsibility and high achievement. *Theory into Practice, 45*, 40–46.

Boaler, J., & Staples, M. (2008). Creating mathematics futures through an equitable teaching approach: The case of Railside school. *Teachers College Record, 110*(3), 608–645.

Bransford, J.D., Brown, A. L., Cocking, R. R. (2000). *How people learn: Brain, mind, experience, and school: Expanded edition*. In R. R. Bransford, J. D. Brown, A. L. Cocking (Eds.), National Research Council.

Carpenter, T. P., Fennema, E., Franke, M. L., Levi, L. W., & Empson, S. B. (2015). *Children's mathematics: Cognitively guided instruction* (2nd ed). Heinemann.

Carpenter, T. P., Franke, M. L., Jacobs, V. R., Fennema, E., & Empson, S. B. (1998). A longitudinal study of invention and understanding in children's multidigit addition and subtraction. *Journal for Research in Mathematics Education, 29*, 3–20. https://doi.org/10.2307/749715

Cobb, P., & Merkel, G. (1989). Thinking strategies: Teaching arithmetic through problem solving. In P. Trafton & A. Shulte (Eds.), *New directions for elementary school mathematics* (pp. 70–81). National Council of Teachers of Mathematics.

Corbin, J., & Strauss, A. (2007). *Basics of qualitative research: Techniques and procedures for developing grounded theory* (3rd ed.). SAGE.

Creswell, J. W. (2013). *Qualitative inquiry & research design: Choosing among five approaches* (3rd ed.). SAGE.

de Villiers, M. (1990). The role and function of proof in mathematics. *Pythagoras, 24*, 17–24.

DeWalt, K. M., & DeWalt, B. R. (2010). *Participant observation: A guide for fieldworkers*. Rowman Altamira.

Gerstenschlager, N. & Strayer, J. (2019). Number talks for statistics and probability. *Mathematics Teaching in the Middle School, 24*(6), 362. https://doi.org/10.5951/mathteacmiddscho.24.6.0362

Hiebert, J., Stigler, J. W., Jacobs, J. K., Givvin, K. B., Garnier, H., Smith, M., Hollingsworth, H., Manaster, A., Wearne, D., & Gallimore, R. (2005). Mathematics teaching in the United States today (and tomorrow): Results from the TIMSS 1999 video study. *Educational Evaluation and Policy Analysis, 27*(2), 111–132.

Humphreys, C., & Parker, R. (2015). *Making number talks matter: Developing mathematical practices and deepening understanding, grades 4–10*. Stenhouse.

Jorgensen, D. L. (1989). *Participant observation: A methodology for human studies*. SAGE.

Kamii, C. (1994). *Young children continue to reinvent arithmetic–3rd grade: Implications of Piaget's theory.* Teachers College Press.

Lannin, J., Ellis, A. B., & Elliot, R. (2011). Developing essential understanding ofmathematics reasoning for teaching mathematics in prekindergarten–Grade 8. NCTM.

Lave, J., & Wenger, E. (1991). *Situated learning: Legitimate peripheral participation.* Cambridge University Press.

Lester, F. K. (2007). *Second handbook of research on mathematics teaching and learning: A project of the National Council of Teachers of Mathematics.* Information Age Publishing.

Matney, G., Lustgarten, A., & Nicholson, T. (2020). Black holes of research on instructional practice: The case of number talks. *Investigations in Mathematics Learning*. https://doi.org/10.1080/19477503.2020.1804273

Miles, M. M., Huberman, A. M., & Saldana, J. (2014). *Qualitative data analysis: A methods sourcebook (*3rd Ed.). SAGE.

Morrow, S. & Smith, M. L. (1995). Constructions of survival and coping by women who have survived childhood sexual abuse. *Journal of Counseling Psychology, 42*(1), 24–33.

Nunes, T., Schliemann, A. D., & Carraher, D. W. (1993). *Street mathematics and school mathematics.* Cambridge University Press.

Parker R., & Humphreys, C. (2018). *Digging deeper: Making number talks matter even more.* Stenhouse.

Parrish, S. (2014). *Number talks: Helping children build mental math and computation strategies*. Math Solutions. (Original work published 2010)

Parrish, S. D. (2011, October). Number talks build numerical reasoning. *Teaching Children Mathematics*, 198–206. https://doi.org/10.5951/teacchilmath.18.3.0198

Parrish, S., & Dominick, A. (2016). *Number talks: Fractions, decimals, and percentages*. Math Solutions.

Pinto, E., & Cañadas, M. C. (2019). Generalizations of third and fifth graders within a functional approach to early algebra. *Mathematics Education Research Journal*. https://doi.org/10.1007/s13394-019-00300-2

Reys, R. E. (1984). Mental computation and estimation: Past, present, and future. *The Elementary School Journal, 84*(5), 546–557.

Russell, S. J., (1999). Mathematical reasoning in the elementary grades. In L.V. Stiff (Ed.), *Developing mathematical reasoning in Grades K–12, 1999 yearbook of the National Council of Teachers of Mathematics* (pp. 1–12). National Council of Teachers of Mathematics.

Siegler, R. S., & Jenkins, E. (1989). *How children discover new strategies*. Erlbaum.

Sowder, J. T. (1988). Making sense of numbers in school mathematics. In G. Leinhardt, R. Putman, & R. Hattrup (Eds.), *Analysis of arithmetic for mathematics*. Erlbaum.

Sowder, J. T. (1992). Making sense of numbers in school mathematics. In G. Leinhardt, R. Putman, & R. Hattrup (Eds.), *Analysis of arithmetic for mathematics teaching* (pp. 1– 51). Lawrence Erlbaum Associates

Sowder, L., & Harel, G. (1998). Types of students' justifications'. *The Mathematics Teacher, 91*(8), 670–675.

Staples, M. E., Bartlo, J., & Thanheiser, E. (2012). Justification as a teaching and learning practice: Its (potential) multifaceted role in middle grades mathematics classrooms. *Journal of Mathematical Behavior, 31*(4), 447–462. https://doi.org/10.1016/j.jmathb.2012.07.001

Strauss, A., & Corbin, J. M. (1990). *Basics of qualitative research: Grounded theory procedures and techniques*. SAGE.

Stylianides, A. J. (2007). Proof and proving in mathematics education. *Journal for Research in Mathematics Education, 38*(3), 289–321. https://doi.org/10.2307/30034869

Sun, K. L., Baldinger, E. E., & Humphreys, C. (2018). Number talks: Gateway to sense making. *The Mathematics Teacher, 112*(1), 48–54.

Torbeyns J., & Verschaffel, L. (2016). Mental computation or standard algorithm? Children's strategy choices on multi-digit subtractions. *European Journal of Psychology of Education, 31*(2), 99–116. https://doi.org/10.1007/sl

Trafton, P. (1978). Estimation and mental computation: Important components of computation. In *1978 NCTM Yearbook*. NCTM.

Turner, E. E., Drake, C., McDuffie, A. R., Aguirre, J., Bartell, T. G., & Foote, M. Q. (2012). Promoting equity in mathematics teacher preparation: A framework for advancing teacher learning of children's multiple mathematics knowledge bases. *Journal of Mathematics Teacher Education, 15*(1), 67–82. https://doi.org/10.1007/s10857-011-9196-6

Verschaffel, L., & DeCorte, E. (1990). Do non-semantic factors also influence the solution process of addition and subtraction word problems? In H. Mandl, E. DeCorte, N. Bennett, & H. E. Friedrich (pp. 415–429). *Learning and instruction, European research in an international context, 2.2*. Pergamon.

Wenger, E. (1998). *Communities of practice: Learning, meaning, and identity*. Cambridge University Press.

Wood, T. (1999). Creating a context for argument in mathematics class. *Journal for Research in Mathematics Education, 30*(2), 171–191. https://doi.org/10.2307/749609

Wood, T., Williams, G., & McNeal, B. (2006). Children's mathematical thinking in different classroom cultures. *Journal for Research in Mathematics Education, 37*(3), 222–255.

Woods, D. (2018). *Developing ambitious mathematics instruction through number talks*. ProQuest Dissertations Publishing.

Woods, D. (2019, April). *Developing elementary students' justification processes through number talks* [Paper presentation]. The 2019 Annual Meeting of the American Educational Research Association. Toronto, Canada.

CHAPTER 11

EPILOGUE

Recognizing the Special Nature of Mathematics Knowledge for Teaching

Patrick M. Jenlink

INTRODUCTION

Importantly, the depth of teachers' mathematical knowledge is proportional to the quality of mathematics teaching presented in the classroom. Teachers' mathematical knowledge influences how and what teachers teach and how and what their students learn (Ball & Bass, 2000; Ball et al., 2008; Barber, 2018; Cuoco et al., 1996; Harel, 1998, 2008a; Harel & Sowder, 2005; Henningsen & Stein, 21997; Hill & Ball, 2004; Hill et al., 2005; Masingila et al., 2018; Olanoff et al., 2009; Stylianides & Ball, 2004). Ball et al. (2008) explain "the mathematical demands of teaching require specialized mathematical knowledge not needed in other settings" (p. 401). Furthermore, "the demands of ... teaching mathematics create the need for ... a body of mathematical knowledge specialized to teaching" (Ball et al., 2008, p. 401). Teachers with high mathematical knowledge for teaching[1] elicit, understand, and build on student thinking; teachers with low mathematical knowledge for teaching struggle more in their teaching and are unable to follow and build on student thinking (Hill & Charalambous, 2012).

Mathematics as the Science of Patterns:
Making the Invisible Visible to Students Through Teaching, pp. 229–243

Habits of Mind—Mental Acts

The habits of mind[2] teachers bring to teaching mathematics (Harel, 1998; Leikin, 2007; Lim & Selden, 2009) reflect the depth of mathematical knowledge and the ability to enact that knowledge in teaching students mathematics. Harel (1998), in his dual assertions on learning and teaching, focuses on the importance of *ways of thinking* and *ways of understanding*, explaining that "*ways of thinking* are ... apparatuses for filtering and interpreting what we intend to teach ... distinguished from *ways of understanding*. A way of understanding is the meaning(s) [one has] for a specific concept (p. 497). Teaching mathematics and leaning mathematics are both subject to Harel's dual assertions. Importantly, *ways of thinking* and *ways of understanding* are linked to habits of mind and the mental acts[3] that teaching and learning mathematics requires. Habits of mind are fundamental to understanding and thinking with respect to mathematical knowledge; therefore, they are fundamental to teaching and learning mathematics, both for the teacher and the student of mathematics.

Harel (2008b) elaborates, defining mathematics as a discipline consisting of two sets of knowledge, *ways of thinking* and *ways of understanding*. The two sets are further defined:

> Mathematics is a union of two sets: The first set is a collection, or structure, of structures consisting of particular axioms, definitions, theorems, proofs, problems, and solutions. This subset consists of all the institutionalized ways of understanding in mathematics throughout history. The second set consists of all the ways of thinking that are characteristics of the mental acts whose products comprise the first set. (p. 490)

The main implication of this definition, as related to teachers preparing to teach mathematics, is that mathematics should be thought of in terms of the constituent elements of mathematics, that is, *ways of understanding* and *ways of thinking*, "not only in terms of the former, as currently is largely the case" (Harel, 2008b, p. 490). The importance here is that as teachers engage in teaching mathematics, the constructs of *ways of understanding* and *ways of thinking* are used to examine idiosyncrasies and difficulties evidenced by students in thinking and problem solving (Harel, 2008a, 2008c).

According to Harel (2008a) a *way of understanding* is a particular cognitive product of a mental act carried out by an individual and a *way of thinking* is a cognitive characteristic of a mental act. As example, when a teacher makes a statement, either orally or in written form on the display board, that statement will become translated by each individual student into a *way of understanding* that depends on her or his experiences with mathematics and the habits of mind and background the student brings to the learning experience. *Ways of thinking*, such as when learning mathematical concepts[4] can be understood in different ways. And, mathematical concepts should be understood in different ways, and it is advantageous to change *ways of understanding* of a mathematical concept in the process of solving problems.

Harel (2008c) explains that teachers, not unlike students, engage in mental acts when teaching mathematics just as students engage in mental acts when solving mathematics problems. Harel further explains, "for any mental act, the *ways of understanding* one produces impact the quality of the *ways of thinking* one forms" (p. 899). Equally importance with respect to understanding mental acts is the converse, namely "for any mental act, the *ways of thinking* one has formed impact the quality of the ways of understanding one produces" (p. 899). For the teacher committed to teaching mathematics, the use of mathematical tasks, which play a critical role in the teaching and learning of mathematics[5] (Henningsen & Stein, 1997), and problem solving engages the student of mathematics in mental acts that draw attention to the habits of mind of the students.[6] The form of mathematical reasoning associated with mathematical tasks and problem solving, that is, the production of proofs for the solution to the problem, may yield wrong or unsubstantiated answers. What is critical is for the teacher of mathematics to draw into specific clarity student thinking that demonstrates students' understanding, difficulties, and misconceptions (Barber, 2019). What is important to recognize is that, as mathematics teacher, one is responsible for engaging the student of mathematics in examining the mathematical reasoning and mental acts used in problem solving or proving to better understand the complexities of mathematical problem and how habits of mind may influence one's solution set (Harel, 2008c). The mathematics teacher's reflection on their own teaching of mathematics is an inherent part of the work of teaching mathematics, with this reflection involving "examining, framing, and attempting to solve the dilemmas of classroom practice; and being aware of and questioning the assumptions and values he or she brings to teaching" (Chapman, 2008, p. 121).

The complexities of teaching mathematics reside, in large part, in the acquisition of mathematical knowledge and proportionally in part in cognition, the mental processes involving knowledge, knowing, and understanding (Karaali, 2019). Preparing teachers of mathematics requires a focus on mathematical cognition or cognitive competence in mathematics, which raises questions about mathematical knowledge, knowing mathematics, and understanding mathematics essential to having the mathematical knowledge required to successfully enter the classroom and teach mathematics, regardless of the grade level. Importantly, Harel's (2008a, 2008b, 2008c) arguments for *ways of understanding* and *ways of thinking* are critical to considerations for how to prepare teachers of mathematics with the cognitive competence to teach mathematics. To further examine *ways of understanding* in relation to teacher cognition and mathematical competence, *ways of understanding* as a subset of Harel's (2008a, 2008b, 2008c) research represents a collection, or structure, of structures consisting of particular axioms, definitions, theorems, proofs, problems, and solutions. This subset consists of all the institutionalized ways of understanding in mathematics throughout history. *Ways of thinking*, as the second subset, consists of all the ways of thinking, which are characteristics of the mental acts whose products comprise the first set.

Mathematical Proficiencies

Making sense of the structure related to *ways of understanding* (Harel, 2008a, 2008b) requires embracing the language of mathematics. An emphasis on the development of mathematical proficiency directly relates to continually developing and evolving the language of mathematics in concert with an understanding of component skills such as concepts, procedures, algorithms, computational. The depth of understanding the language of mathematics correlates to the learning of mathematics and the ways of thinking mathematically that are essential to learning. The teacher of mathematics is responsible for ensuring that, through her or his teaching, the students learning mathematics are equally focused on learning and understanding the language of mathematics and procuring a mathematics vocabulary (Riccomini et al., 2015). Equally important is a focus on understanding of component skills such as concepts, procedures, algorithms, computation, and problem solving (Riccomini et al., 2008).

The ability to engage in mathematical tasks and problem solving,[7] deriving proofs of the solutions to mathematical problems, requires a level of familiarity and comfort with the language of mathematics (i.e., learning and using the terms associated with mathematics). The mental acts and habits of mind related to how teachers and students engage in mathematical thinking and understanding are dependent on an increasingly sophisticated level of mathematics language understanding. The language and understanding of mathematical terms is quite literally embedded in the cognitive reasoning required to think and understand mathematically. To this point, the National Research Council (2001) described emphasis on language development as critical to the student of mathematics having the "capacity for logical thought, reflection, explanation, and justification" (p. 116). As Riccomini et al. (2015) further explain the importance of knowing and understanding mathematical language, in mathematics courses "students often only develop surface-level understanding of the material, and without opportunities for further engagement students will not gain the desired deep understanding necessary for mathematical reasoning and communication" (p. 239).

Teaching mathematics and students learning mathematics rests, in part, on an increasingly higher level of understanding of the mathematics language, if proficiency in mathematics is to be attained. Proficiency in mathematical tasks depends on critical component skills such as understanding mathematical concepts, procedures, algorithms, computation, problem solving, and language. The importance of understanding mathematics language lies in the teacher of mathematics engaging her or his students in a mathematics discourse[8] necessary to teaching and learning mathematics. Equally important, understanding mathematics language is directly related to mathematical task knowledge (Barber, 2018).

The cognitive demands associated with teaching, for teachers of mathematics as well as students learning of mathematics, present an important consideration for designing mathematical tasks. A proficiency in mathematics language correlates with the cognitive reasoning and *ways of thinking* mathematically (Harel, 2008a, 2008b). The thinking processes required to design a mathematical task as well as the thinking processes required of students in solving the task place cognitive demands on both the teacher and student (Henningsen & Stein, 1997). Recognizing the importance of cognitive reasoning aligned with teaching and learning mathematics is crucial for teaching mathematics. Henningsen and Stein focus attention specifically on the range of cognitive demands that teaching mathematics through mathematical tasks requires, noting the range of demands from memorization, use of procedures and algorithms (with or without attention to concepts, understanding, or meaning) to complex thinking and cognitive reasoning, strategies typical to problem solving and mathematical task completion (e.g., conjecturing, justifying, interpreting).

A proficient mathematics teacher must, as Lester (2013) argues, be proficient in using a variety of representations and posses a sound grasp of how to recognize and construct patterns of inference. Moreover, proficiency as a mathematics teacher "recognizes the importance of intuition in successful problem solving" (p. 249). Proficiencies for teaching mathematics align, or should align, with the proficiencies for preparing mathematics teachers, in each case, as Lester explains "such proficiencies must take into account the assumptions that are being made about the nature of mathematics learning and instruction, as well as about instructional goals. For example, a proponent of direct instruction" (p. 259). As noted in the previous discussion, if problem solving is the means through which mathematical concepts, processes, and procedures are learned either as mathematics teacher or student in a mathematics classroom, then a different set of proficiencies may be called for with respect to designing problem solving and mathematical tasks for mathematics teacher preparation versus teaching mathematics through the use of problem solving and mathematical tasks. As Lester explains, for example, in the school classroom where mathematics instruction is carried out "the teacher for whom problem solving is a means, would likely need to be very proficient at listening to and observing students as they work on mathematical tasks" (p. 260). Stated more directly and with consideration of how to include problem solving mathematical tasks "in a mathematical-proficiencies-for-teaching framework should be done in view of the assumptions the teacher makes about the nature of mathematics learning and the goals of instruction" (p. 260).

A mathematics proficient must have mastery of how to design and select appropriate tasks in coordination with making sense of and taking appropriate actions after listening to and observing students as they engage in problem solving and work through the mathematics task to produce a solution and proof. In concert, this requires designing tasks that are appropriately problematic for students and developing pedagogical and assessment strategies that enable the mathematics teacher to pay close attention to and being familiar with the methods students use to problem solve and derive a solution and proof. Mathematics teachers using problem solving and mathematical tasks will need to be proficient at mindfulness of the students' work and when to take appropriate actions. Affording students the appropriate opportunities to engage in problem solving and completing mathematical tasks requires a classroom atmosphere that is conductive to creative exploration, sharing, and oral communication to discuss the selected strategies and cognitive reasoning employed in deriving a solution and proof for the assigned problem (Lester, 2013).

Mathematical-Task Knowledge

Mathematical tasks also play a critical role in the teaching and learning of mathematics (Barber, 2018; Baumert et al., 2010; Chamberlin & Zawojewski, 2006; Crespo, 2003; Doerr, 2006; Goldenberg, 1996; Henningsen & Stein, 1997; Stein et al., 1996). Primarily, the mathematical knowledge for teaching is concerned with tasks and problem solving[9] necessary to student learning of mathematics. "Central to these tasks is knowledge of common student conceptions and misconceptions about particular mathematical content" (Ball et al., p. 401). Mathematical tasks are important instructional considerations for developing student capacity for mathematical thinking and reasoning (Stein et al., 1996). As Masingila et al. (2018) explain, there is a differential between high-level and low-level tasks. "High-level tasks engage students in making connections among mathematical ideas and doing mathematics—making and testing conjectures, justifying solutions, solving problems for which a procedure is not known—while low-level tasks involve memorization and procedures that are not connected to conceptual ideas" (p. 533). Specific to mathematical tasks, mathematical-task knowledge is the knowledge teachers need to select, develop, and/or modify high-quality mathematical tasks. Importantly, teaching mathematics requires an understanding, on the part of teacher, of mathematical-task knowledge and how mathematical tasks are designed and used to increase student learning by supporting students' conceptual understanding, cognitive reasoning, and by experiencing rich content (Barber, 2018).

Chapman (2013) explains that mathematical-task knowledge for teaching "is multidimensional and thus likely to be challenging for a teacher to construct without meaningful intervention to build on her or his initial sense-making of tasks" (p. 1). Further explaining the importance of mathematical-task knowledge, Chapman notes that mathematical-task knowledge focuses on "knowledge teachers need in order to (a) select and develop tasks to promote students' conceptual understanding of mathematics, support their development of mathematical thinking, and capture their interest and curiosity and (b) optimize the learning potential of such tasks" (p. 1). When considering *ways of understanding* and *ways of thinking* (Harel, 2008a, 2008b, 2008c), mathematical-task knowledge and the incorporation of mathematical tasks in teaching presents opportunity for teaching mathematics that is focused on developing conceptual understanding and high-level thinking.

Designing mathematical tasks for teaching is concerned with how teachers think about, select, and develop tasks, which in turn determines how the teacher uses the tasks in her or his classroom (Barber, 2018; Charalambous, 2010; Nyman, 2016). Chapman (2013) explains that the teacher engaged in using mathematical tasks must understand what constitutes a worthwhile task for learning mathematics, which is directly correlated to the level of mathematical task knowledge the teacher has developed. Raghubar et al. (2010) draw attention to the importance of working memory in mathematical tasks, noting that working memory refers "to a mental workspace, involved in controlling, regulating, and actively maintaining relevant information to accomplish complex cognitive tasks (e.g. mathematical processing)" (p. 110). Designing mathematical tasks requires consideration of the cognitive reasoning and the cognitive tasks associated with completing a mathematical task.

Equally important, using mathematical tasks requires the "ability to identify, select, and create tasks that are rich mathematically in terms of content" (Chapman, 2013, p. 1). Designing mathematical tasks that are rich mathematically requires of the teacher the "knowledge of levels of cognitive demands of tasks" (Chapman, 2013, p. 2). Understanding the diversity of learner, the teacher of mathematics needs to have "knowledge of students' understandings, interests, and experiences and the range of ways that diverse students learn mathematics" (Chapman, 2013, p. 2). As well, the teacher of mathematics needs "understanding of how the tasks teachers select and how they use them influence how students come to make sense of mathematics and of doing and applying mathematics" (Chapman, 2013, p. 2). Finally, drawing from the National Council of Teachers of Mathematics [NCTM] standards, Chapman (2013) focuses attention on the importance of having task knowledge of "what aspects of a task to highlight, how to organize … the work of the students, … questions to ask to challenge those with varied levels of expertise, and how to support students without taking over the process of thinking for them" (NCTM, 2000, p. 19).

Importantly, mathematical-tasks may be used to examine patterns—pattern exploration tasks—in a variety of contexts including numeric, geometric, pictorial, and through the application of different approaches. Pattern exploration tasks may be used to analyze "difficulties presented by students, strategies emerging from their work, and the impact of visual contexts in generalization" (Barbosa, et al., 2012, p. 290). Cognitive reasoning patterns used in problem solving are detectable in different contexts using and understanding symbols and variables that represent patterns and generalizing, which are significant components of the mathematical-task. Returning to Harel's (2008a, 2008b, 2008c) focus on *ways of understanding* and *ways of thinking*, there are invisible patterns in the cognitive reasoning of mathematical tasks that become visible through pattern exploration. The mathematics teacher, by engaging students in examining the solution or proof derived from the mathematical task, is able to draw the patterns out and engage the student in understanding how she or he arrived at the solution or proof. The idea of visual reasoning approaches for supporting problem solving and mathematical learning presents an opportunity to make links between visual[10] and analytical thought (Presmeg, 1986, 2006).

Mathematical tasks play a critical role in the teaching and learning of mathematics. Worthwhile tasks give students the chance to solidify and extend what they know and to stimulate mathematics learning (NCTM, 2010). Mathematical-task knowledge needed for teaching via mathematical tasks spans across several knowledge domains including common content knowledge, specialized content knowledge, knowledge of content and students, and knowledge of content and teaching (Ball et al., 2008; Hill et al., 2008; Hill et al., 2004). Since mathematical-task knowledge can include how teachers understand, select, and modify tasks (specialized content knowledge), anticipating and interpreting student thinking (knowledge of content and students), and how teachers' use of tasks influence student learning (knowledge of content and teaching), mathematical-task knowledge for teachers is complex and of critical importance to teachers' learning to teach mathematics as well as teachers' ability to teach mathematics to diverse students in the classroom. Importantly, how teachers think about, select, and develop tasks will determine how they use the tasks in their classroom.

As mathematics teachers engage in *ways of thinking* and *ways of understanding* (Harel, 2008a, 2008b, 2008c) associated with problem solving and task completion, designing mathematical tasks requires of mathematics teachers a high level of cognitive reasoning aligned with mathematical knowledge. While sound mathematical tasks can help develop mathematical-task knowledge, "such engagement may help them [teachers] to understand an isolated aspect of this knowledge, but not all of the components that are necessary for meaningful implementation" (Chapman, 2013, p. 6). Developing mathematical-task knowledge is ongoing cognitive process. Lappan and Briars (1995) are instructive when they state, "There is no decision that teachers make that has a greater impact on students' opportunities to learn, and on their perceptions about what mathematics is, than the selection or creation of the tasks" (p. 138).

The mathematical-task knowledge that a teacher of mathematics has reflects directly on their ability to identify, select, and create tasks that ensure that the *ways of understanding* and *ways of thinking* (Harel, 2008a) mathematically provide students with a deep understanding of mathematics. A well-designed mathematical task for learning mathematics creates a cognitive means for the student of mathematics to visualize,[11] discuss, and connect to important mathematical concepts. Implementing mathematical tasks in a way that supports students' *ways of thinking* and *ways of understanding* (Harel, 2008a, 2008b, 2008c) mathematics, without reducing the level of cognitive demand or taking over the most challenging aspects of the task, provides authentic learning experiences for students of mathematics (Barber, 2018; Chapman, 2013).

CONCLUSIONS

The mathematical knowledge for teaching necessary to enter the classroom and teach students mathematics is a mathematics knowledge defined in large part by its complexities and at the same time challenging due to the diversity of habits of mind that individuals bring to mathematics teacher preparation and later enter the classroom with to teach. Germane to this point, Stein et al. (2008) state, "At the heart of the challenge associated with student-centered practice is the need to strike an appropriate balance between giving students authority over mathematical work and ensuring that this work is held accountable to the discipline" (p. 332). The nature of mathematics knowledge for teaching emphasized herein has focused on *ways of understanding* and *ways of thinking* aligned with mathematics as a discipline.

The mathematics teacher and the student are both subject to these two subsets of mathematics knowledge: the teacher in order to understand how to design and facilitate student learning of mathematics and the student in order to understand how to engage in problem solving in mathematics. Importantly, as Stein et al. (2008) explain there is need, on the part of the mathematics teacher, to nurture and ensure a level of "mathematics authority" in the students they teach, providing mathematics problem solving and tasks that provide "opportunity for students to publicly engage in the solving of real mathematical problems, actively grapple with various strategies and representations that they devise to make headway on the problems, and judge the validity and efficacy of their own approaches themselves" (p. 332). Drawing on this reasoning, there is equal need on the part of the teacher learning mathematics to have mathematics teacher educators who instill a similar level of "mathematics authority" in the mathematics teacher.

The habits of mind, mental acts, and mathematical proficiencies necessary to teaching mathematics in todays' educational settings speak to the importance of a high level of "mathematics authority" in *ways of understanding* and *ways of thinking* mathematically. Equally important, a high level of "mathematical authority" in understanding the language, logic of mathematical patterns, and cognitive competence required of successful mathematics teachers stand without question.

NOTES

1. Mathematical Knowledge for Teaching (MKT) is the mathematical knowledge needed to perform the recurrent tasks of teaching mathematics. Specifically, MKT is essential in recognizing, understanding, and responding to the mathematical problems and tasks encountered in teaching the subject (Ball et al., 2008; Phelps & Howell, 2016).
2. Habits of mind as considered in mathematical context are guiding mathematical principles for individual mathematical reasoning (Cuoco et al., 1996). Habits of mind are associated with concepts and principles incorporated in one's repertoire so that they are applied successfully in an appropriate problem-solving situation (Leikin, 2007).
3. Harel and Sowder (2005) use the term "mental act" to denote any internal mental action, such as interpreting, inferring, explaining, or searching. One single cognitive product of a mental act (or acts) in one given situation is termed a *way of understanding*. If, in observing a student, a particular characteristic is found to be repeatedly associated with a given mental act, then Harel and Sowder term that a *way of thinking*.
4. For the teacher of mathematics, it is important to understand that typically, students' definitions of concepts are not mathematical, even if the concepts were defined to them mathematically. Understanding the notion of *mathematical definition* and appreciating the role and value of mathematical defi-

nitions in problem solving and proving is a developmental process, which is not achieved for most students until adulthood (Harel, 2008a).

5. Masingila et al. (2018) explain the importance of tasks: "High-level tasks engage students in making connections among mathematical ideas and doing mathematics—making and testing conjectures, justifying solutions, solving problems for which a procedure is not known—while low-level tasks involve memorization and procedures that are not connected to conceptual ideas" (p. 533).
6. To the point, Harel (2008c) explains: "the proofs children produce to prove assertions and account for phenomena in everyday life impact the kind and robustness of the proof schemes they form. Proofs ... are ways of understanding associated with the mental act of proving, and proof schemes are ways of thinking associated with the same act" (p. 899).
7. Speaking to the meaning of problem solving and breadth of knowledge and proficiencies required of the mathematics teacher, Lester and Kehle (2013) describe problem solving as: "Successful problem solving involves coordinating previous experiences, knowledge, familiar representations and patterns of inference, and intuition in an effort to generate new representations and related patterns of inference that resolve some tension or ambiguity (i.e., lack of meaningful representations and supporting inferential moves) that prompted the original problem-solving activity" (p. 510).
8. While the mathematics teacher telling her or his students mathematical information is easier, engaging students in learning through mathematics task and problem solving results in deeper and more meaningful learning. Discourse is important to student learning of mathematics. When students engage in mathematical tasks and problem solving they learn mathematics differently than they do when a mathematics teacher presents mathematical information orally or via posting it on a board. Marcus and Fey (2003) explain that "even the clearest teacher explanations leave many students with incomplete understanding and shaky confidence. Ideas that are forged by hard thought and tested in discourse with other students and teachers are much more likely to last and be useful" (p. 61).
9. Teaching mathematics using problem solving and mathematical tasks presents several challenges. The mathematics teacher is responsible for selecting a mathematic problem, designing the task, and facilitating the students as they engage with the tasks that are appropriate for the mathematical problem solving that she or he wants the students to engage in. This involves choosing high-level tasks and implementing the tasks in ways that maintain the cognitive demand of the tasks (Henningsen & Stein, 1997; Masingila et al., 2018).
10. Barwise and Etchemendy (1991) are instructive in understanding the connection between visual representations, visual reasoning and problem solving, explaining that "visual forms of representation can be important ... as legitimate elements of mathematical proofs" (p. 9).
11. Pólya (1957), in his book *How to Solve It,* speaks to the importance of visualizing a problem or task, "visualize the problem as a whole as clearly and as vividly as you can. Do not concern yourself with the details for the moment"

(p. 33). Visualizing opens the problem to the mind and presents an opportunity to "isolate the principal parts of a 'problem to prove,'" which are the "hypothesis and the conclusions" whereas the "unknown, the data, and the conditions are the principal parts of a 'problem to find'" (p. 33). Clarification of the problem or mathematical task is important before engaging in discussion and finding the solution.

REFERENCES

Ball, D. L., & Bass, H. (2000). Interweaving content and pedagogy in teaching and learning to teach: Knowing and using mathematics. In J. Boaler (Ed.), *Multiple perspectives on the teaching and learning of mathematics* (pp. 83–104). Ablex.

Ball, D. L., Thames, M. H., & Phelps, G. (2008). Content knowledge for teaching: What makes it special? *Journal of Teacher Education, 59,* 389–407. https://doi.org/10.1177/0022487108324554

Barber, K. (2018). Developing teachers' mathematical-task knowledge and practice through lesson study. *International Journal for Lesson and Learning Studies, 7*(2), 136–149. https://doi.org/10.1108/IJLLS-09-2017-0042

Barbosa, A., Vale, I., & Palhares, P. (2012). Pattern tasks: Thinking processes used by 6th grade students. *Revista Latinoamericana de Investigación en Matemática Educativa, 15*(3), 273–293.

Barwise, J., & Etchemendy, J. (1991). Visual information and valid reasoning. In W. Zimmermann & S. Cunningham (Eds.), *Visualization in teaching and learning mathematics* (pp. 9–24) Mathematical Association of America.

Baumert, J., Kunter, M., Blum, W., Brunner, M., Voss, T., Jordan, A., Klusmann, U., Krauss, S., Neubrand, M., & Tsai, Y.-M. (2010). Teachers' mathematical knowledge, cognitive activation in the classroom, and student progress. *American Educational Research Journal, 47*(1), 133–180. https://doi.org/10.3102/0002831209345157

Chapman, O. (2008). Mathematics teacher educators' learning from research on their instructional practices: A cognitive perspective. In B. Jaworski & T. Wood (Eds.), *The international handbook of mathematics teacher education: The mathematics teacher educator as a developing professional* (Vol. 4, pp. 115–134). Sense.

Chapman, O. (2013). Mathematical-task knowledge for teaching. *Journal of Mathematics Teacher Education, 16,* 1–16. https://doi.org/10.1007/s10857-013-9234-7

Charalambous, C. (2010). Mathematical knowledge for teaching and task unfolding: An exploratory study. *The Elementary School Journal, 110*(3), 247–278. https://doi.org/10.1086/648978

Chamberlin, M. T., & Zawojewski, J. (2006). A worthwhile mathematical task for students and their teachers. *Mathematics Teaching in the Middle School, 12*(2), 82–87.

Crespo, S. (2003). Learning to post mathematical problems: Exploring changes in preservice teachers' practices. *Educational Studies in Mathematics, 52*(3), 243–270. https://doi.org/10.1023/A:1024364304664

Cuoco, A., Goldenberg, E. P., & Mark, J. (1996). Habits of mind: An organizing principle for mathematics curricula. *Journal of Mathematical Behavior, 15*, 375–402. https://doi.org/10.1016/S0732-3123(96)90023-1

Doerr, H. M. (2006). Examining the tasks of teaching when using students' mathematical thinking. *Educational Studies in Mathematics, 62(*3), 3–24. http://dx.doi.org/10.1007/s10649-006-4437-9

Goldenberg, P. (1996). "Habits of mind" as an organizer for the curriculum. *Journal of Education, 178*, 13–34. https://doi.org/10.1177/002205749617800102

Harel, G. (1998). Two dual assertions: the first on learning and the second on teaching (or vice versa). *The American Mathematical Monthly, 105*(6), 497–507. https://doi.org/10.1080/00029890.1998.12004918

Harel, G. (2008a). What is mathematics? A pedagogical answer to a philosophical question. In B. Gold & R. A. Simons (Eds.), *Proof and other dilemmas: Mathematics and philosophy* (pp. 265–290). MAA Press.

Harel, G. (2008b). DNR perspective on mathematics curriculum and instruction, Part I: focus on proving. *ZDM Mathematics Education, 40*, 487–500. https://doi.org/10.1007/s11858-008-0104-1

Harel, G. (2008c). A DNR perspective on mathematics curriculum and instruction. Part II: with reference to teacher's knowledge base. *ZDM Mathematics Education, 40*, 893–907. https://doi.org/10.1007/s11858-008-0146-4

Harel, G., & Sowder, L. (2005). Advanced mathematical-thinking at any age: Its nature and its development. *Mathematical Thinking and Learning, 7*(1), 27–50. https://doi.org/10.1207/s15327833mtl0701_3

Henningsen, J., & Stein, M. K. (1997). Mathematical tasks and student cognition: Classroom-based factors that support and inhibit high-level mathematical thinking and reasoning . *Journal for Research in Mathematics Education, 28*(5), 524–549. https://doi.org/10.2307/749690

Hill, H. C., & Ball, D. L. (2004). Learning mathematics for teaching: Results from California's Mathematics Professional Development Institutes. *Journal of Research in Mathematics Education, 35*, 330–351. https://doi.org/10.2307/30034819

Hill, H. C., & Charalambous, C. Y. (2012). Teacher knowledge, curriculum materials, and quality of instruction: Lessons learned and open issues. *Journal of Curriculum Studies, 44*(4), 559–576. https://doi.org/10.1080/00220272.2012.716978

Hill, H. C., Blunk, M. L., Charalambous, C. Y., Lewis, J. M., Phelps, G. C., Sleep, L., & Ball, D. L. (2008). Mathematical knowledge for teaching and the mathematical quality of Instruction: An exploratory study. *Cognition and Instruction, 26*(4), 430–511. https://doi.org/10.1080/07370000802177235

Hill, H. C., Rowan, B., & Ball, D. L. (2005). Effects of teachers' mathematical knowledge for teaching on student achievement. *American Educational Research Journal, 42*(2), 371–406. https://doi.org/10.3102/00028312042002371

Hill, H., Schilling, S., & Ball, D. (2004). Developing measures of teachers' mathematics knowledge for teaching. *The Elementary School Journal, 105*(1), 11–30. https://doi.org/10.1086/428763

Karaali G. (2019). On mathematical ways of knowing: Musings of a humanistic mathematician. In M. Danesi (Ed.), *Interdisciplinary perspectives on math cognition* (pp. 321–332). Springer Nature. https://doi.org/10.1007/978-3-030-22537-7_19

Lappan, G., & Briars, D. (1995). How should mathematics be taught? In I. M. Carl (Ed.), *Prospects for school mathematics: Seventy-five years of progress* (pp. 131–156). Reston, VA: National Council of Teachers of Mathematics.

Leikin, R. (2007). Habits of mind associated with advanced mathematical thinking and solution spaces of mathematical tasks. In D. Pitta-Pantazi, & G. Philippou (Eds.), *Proceedings of the Fifth Conference of the European Society for Research in Mathematics Education—CERME-5* (pp. 2330–2339). Larnaca, Cyprus. http://www.mathematik.uni-dortmund.de/~erme/CERME5b/WG14.pdf#page

Lester, F. K., Jr. (2013). Thoughts about research on mathematical problem-solving instruction. *The Mathematics Enthusiast, 10*(1-2), 245–278. https://scholarworks.umt.edu/tme/vol10/iss1/12

Lester, F. K., & Kehle, P. (2003). From problem solving to modeling: The evolution of thinking about research on complex mathematical activity. In R. Lesh & H. M. Doerr (Eds.), *Beyond constructivism: Models and modeling perspectives on mathematics problem solving, learning, and teaching* (pp. 501–517). Erlbaum.

Lim, K., & Selden, A. (2009). Mathematical habits of mind. In S. L. Swars, D. W. Stinson, & S. Lemons-Smith (Eds.), *Proceedings of the 31st annual meeting of the North American Chapter of the International Group for the Psychology of Mathematics Education* (pp. 1576–1583). Georgia State University. https://owl.purdue.edu/owl/research_and_citation/apa_style/apa_formatting_and_style_guide/reference_list_books.html

Marcus, R., & Fey, J. T. (2003). Selecting quality tasks for problem-based teaching. In H. L. Schoen & R. I. Charles (Eds.), *Teaching mathematics through problem solving: Grades 6–12* (pp. 55–67). National Council of Teachers of Mathematics.

Masingila, J. O., Olanoff, D., & Kimani, P. M. (2018). Mathematical knowledge for teaching teachers: Knowledge used and developed by mathematics teacher educators in learning to teach via problem solving. *Journal of Mathematics Teacher Education, 21*, 429–450. https://doi.org/10.1007/s10857-017-9389-8

National Council of Teachers of Mathematics. (2000). *Principles and standards for school mathematics*.

National Research Council. (2001). *Adding it up: Helping children learn mathematics*. J. Kilpatrick, J. Swafford, & B. Findell (Eds.), *Mathematics Learning Study Committee Center for Education, Division of Behavioral and Social Sciences and Education*. National Academies Press.

National Council of Teachers of Mathematics. (2010, April). *Research brief: Problem solving*. https://www.nctm.org/Research-and-Advocacy/Research-Brief-and-Clips/Problem-Solving/

Nyman, R. (2016). What makes a mathematical task interesting? *Educational Research and Review, 11*(16), 1509–1520. https://doi.org/10.5897/ERR2016.2919

Olanoff, D. E, Kimani, P. M., & Masingila, J. O. (2009). Learning to teach via problem solving and supporting pre-service teachers in learning via problem solving. In S. L. Swars, D. W. Stinson, & S. Lemons-Smith (Eds.), *Proceedings of the 31st annual meeting of the North American Chapter of the International Group for the Psychology of Mathematics Education* (pp. 1299–1307). Georgia State University.

Phelps, G., & Howell, H. (2016). Assessing mathematical knowledge for teaching: The role of teaching context. *The Mathematics Enthusiast, 13*(1–2), 52–70. https://scholarworks.umt.edu/tme/vol13/iss1/1

Pólya, G. (1957). *How to solve it* (2nd ed.). Princeton University Press (1st edition, 1945).

Presmeg, N. (1986). Visualization and mathematical giftedness. *Educational Studies in Mathematics 17*(3), 297–311. https://doi.org/10.1007/BF00305075

Presmeg, N. (2006). Research on visualization in learning and teaching mathematics: Emergence from psychology. In A. Gutiérrez, & P. Boero (Eds.), *Handbook of research on the psychology of mathematics education* (pp. 205–235). Sense.

Raghubar, K. P., Barnes, M. A., & Hecht, S. A. (2010). Working memory and mathematics: A review of developmental, individual difference, and cognitive approaches. *Learning and Individual Differences, 20*, 110–122. https://doi.org/10.1016/j.lindif.2009.10.005

Riccomini, P. J., Sanders, S., & Jones, J. (2008). The key to enhancing students' mathematical vocabulary knowledge. *Journal on School Educational Technology, 4*(1), 1–7. https://doi.org/10.26634/jsch.4.1.627

Riccomini, P. J., Smith, G. W., Hughes, E. M., & Fries, K. F. (2015). The language of mathematics: The importance of teaching and learning mathematical vocabulary. *Reading & Writing Quarterly, 31*(3), 235–252. https://doi.org/10.1080/10573569.2015.1030995

Stein, M. K., Engle, R. A., Smith, M. S., & Hughes, E. K. (2008). Orchestrating productive mathematical discussions: Five practices for helping teachers move beyond show and tell. *Mathematical Thinking and Learning, 10*(4), 313–340 http://dx.doi.org/10.1080/10986060802229675

Stein, M. K., Grover, B. W., & Henningsen M. A. (1996). Building student capacity for mathematical thinking and reasoning: An analysis of mathematical tasks used in reform classrooms. *American Educational Research Journal, 33*(2), 455–488. https://doi.org/10.3102/00028312033002455

Stylianides, A. J., & Ball, D. L. (2004, April). *Studying the mathematical knowledge needed for teaching: The case of teachers' knowledge of reasoning and proof* Paper presentation]. The 2004 Annual Meeting of the American Educational Research Association, San Diego, CA.

ABOUT THE EDITOR AND AUTHORS

ABOUT THE EDITOR

Patrick M. Jenlink is Regents Professor, College of Education, Stephen F. Austin State University. He has served as E. J. Campbell Endowed Chair in Educational Leadership, doctoral program coordinator, director of the Educational Research Center, and department chair during his tenure. His experience over the past four decades as an educator includes STEM teaching in public schools as well as serving as a building administrator, school district superintendent, professor at Western Michigan University, an evaluator and research consultant on National Science Foundation funded Statewide Systemic STEM initiative in Michigan, senior researcher on funded STEM initiatives in Oklahoma with NASA and Oklahoma State University, and senior researcher in grant funded STEM initiative in Texas, the *Texas State Middle Level Mathematics Project*. Dr. Jenlink's research interests include STEM literacy, mathematics teaching and learning, meta-disciplinarity and pedagogy, educator preparation, STEM architecture, and leadership for STEM innovation. He has edited and/or authored 19 books and authored over 95 book chapters. As well, he has authored and published over 175 peer-refereed articles and over 200 hundred peer-refereed conference papers. His most recent books include *STEM Teaching: An Interdisciplinary Approach* (Rowman & Littlefield Publishing Group), *The Language of Mathematics: How the Teacher's Knowledge of Mathematics Affects*

Instruction (Rowman & Littlefield Publishing Group), *Multimedia Learning Theory and its Implications for Teaching and Learning* (Rowman & Littlefield Publishing Group), *Dewey Studies Handbook* (Brill/Sense). Current book projects include *STEM Teacher Preparation and Practice for the 21st Century: Research-Based Insights*.

ABOUT THE AUTHORS

Jair J. Aguilar holds a PhD in STEM education with a concentration in Mathematics from the University of Texas, Austin. Prior to earning his doctoral degree, he earned a Bachelor of Science in Engineering with a concentration in Mathematics and Computer Science from the Monterrey Tech., a Master of Education from Monterrey University, and a MBA from the Regiomontana University. Dr. Aguilar has over 20 years of teaching experience ranging from middle school to higher education. His research focuses on how students solve open-ended, nonroutine mathematical tasks and the intersections between content knowledge, technology, beliefs, and self-efficacy. Dr. Aguilar's recent research projects include studying the learning of fractions in at the elementary school level, the impact of Mixed-Reality Simulations on preservice elementary mathematics teachers ability to ask higher-order questions, and strategies for eliciting students' thinking.

Didem Akyüz is an Associate Professor at the Department of Mathematics and Science Education lecturer in the Faculty of Education at Middle East Technical University. Her studies mainly focus on technology in mathematics education, teacher practices, beliefs, goals, and mathematics teacher education. Dr. Akyüz is also giving lectures to the graduate and undergraduate elementary mathematics education students and conducting research based on the practices of the teachers and teacher candidates.

Christina Areizaga Barbieri is an Assistant Professor at the University of Delaware's School of Education. Dr. Barbieri's research program centers on instruction for students who have difficulties with mathematics. Specifically, her work focuses on the evaluation and application of learning principles to improve mathematical competencies, especially for students with a history of low mathematics performance. Dr. Barbieri also considers the role of motivation and attitudes towards mathematics in student learning.

Julie L. Booth is a Professor of STEM Education and Educational Psychology at Temple University and Deputy Dean of Academic and Faculty Affairs for Temple's College of Education and Human Development. Her work focuses on translation between cognitive science and education, particularly understanding how students learn math and how instruction can be designed to improve students' math learning.

Kathryn J. deVries is a third-year graduate student in the doctoral program in the College of Education and Human Development at Temple University. She currently works as a Graduate Research Assistant and previously taught undergraduate and graduate classes. She is a licensed professional counselor who is certified in the Neurosequential Model of Therapeutics and specializes in trauma. This clinical experience has led her to a research interest in the effects of adversity on math performance. Kathryn is proposing a dissertation using the Neurovisceral Integration Model as a framework for understanding how adversity effects student's math outcomes. Doctorate expected Fall 2021.

Viktor Freiman is a full professor in mathematics education at the Faculté des sciences de l'éducation de l'Université de Moncton, Canada. His research interests are focused on actualization of full potential of all students based on creativity, use of technology, innovation, and interdisciplinary collaboration. Coeditor of Springer Book Series Mathematics Education in the Digital Era, he recently published two coedited books on *Creativity and Technology in Mathematics Education* and *Computations and Computing Devices in Mathematics Education Before the Advent of Electronic Calculators*. He is a president-elect of the International Group for Mathematical Creativity and Giftedness. He is also a Director of CompeTI.CA (ICT Competencies in the Atlantic Canada) Partnership Network.

Emunah Mager Garfield, MEd, is a second-year doctoral student in the College of Education and Human Development at Temple University. Her research interests broadly include executive functioning skills, especially as they relate to the social-emotional realm and academic success. She expects to earn her PhD in 2024.

Åsmund Lillevik Gjære is a university lecturer in mathematics and a PhD candidate in educational science at the University of Stavanger, Norway. His research interests include developmental education and the development of mathematical noticing abilities in elementary students, as well as practice-based teacher training practices.

Tracy Goodson-Espy is a Professor in the Department of Curriculum and Instruction at Appalachian State University. Tracy is interested in STEM teacher education at elementary and secondary level and is responsible for teaching quantitative and qualitative research courses in the Doctoral Program.

Sheri Johnson is a member of the Upper School Faculty at The Mount Vernon School, a leading private school of inquiry, innovation, and impact. Her interest lies in how students learn mathematics and statistics and how they are able to apply their knowledge to solve real-world problems. Her engineering degree and background in healthcare systems management complement her recent study of statistics and mathematics education at the University of Georgia. She is a coauthor of the recently released Pre-K–12 Guidelines for Assessment and Instruction in Statistics Education (GAISE II) which is endorsed by the American Statistical Association (ASA) and the National Council of Teachers of Mathematics (NCTM).

Rui Kang is Professor of Secondary Education (6–12) at Georgia College & State University (GCSU). Her scholarship focuses on mathematics teaching and learning, STEM education, and teacher preparation and professional development. Her recent publications appear in journals such as *Mathematics Teacher*, *International Journal of Science and Mathematics Education*, *Teacher Education & Practice*, *The Journal of Social Studies Research*, and *School Science and Mathematics*. She has been an editorial board member of the journal, *Learning for Democracy,* since 2007. She was the Program Chair of the Democratic Citizenship in Education Special Interest Group of the American Education Research Association (AERA) from 2016 to 2018. She has taught high school mathematics and holds a clear renewable teaching certificate in secondary mathematics in the state of Georgia.

Courtney Glavich Mayakis is a former Middle Grades mathematics teacher, and currently an Assistant Professor and Assessment Coordinator at Campbell University. She holds a Bachelor's degree in Elementary Education from North Carolina State University, and a Master's degree in Middle Grades Education from the University of North Carolina at Charlotte. She received her doctoral degree at the University of North Carolina at Charlotte, in Curriculum and Instruction. Her research interests include preservice teacher dispositions, the role of field placements in teacher education programs, and the quality and rigor of STEM pedagogy and content courses in pre-service teacher education in the primary grades.

Diana L. Moss is an Assistant Professor in Mathematics Education in the Department of Curriculum and Instruction at Appalachian State University. She teaches junior and senior undergraduate courses in mathematics content and pedagogy and is a member of the graduate faculty. Her teaching and research interests include children's mathematical thinking and teacher education. Her work involves using design research in classroom settings to support learning through the perspective of social constructivism.

Christopher Nazelli is a Senior Lecturer in the Department of Mathematics at Wayne State University and the Director of Curriculum and Instruction for its Center for Excellence and Equity in Mathematics. In addition to his university teaching, Chris serves within the WSU Math Corps, a nationally recognized program that, each summer, helps over 400 Detroit middle and high school students to realize their own unique greatness. His research interests include preservice teacher preparation, professional development, and lesson study.

Kristie J. Newton is an Associate Professor in the Department of Teaching and Learning at Temple University. Her research focuses on the development of mathematical knowledge, especially related to fractions and algebra. She has explored mathematical thinking across a range of groups, from struggling learners to experts, and is also interested in how this knowledge is related to other significant factors, such as motivation and instruction. Kristie coedited a handbook and authored two books.

Sinem Sözen Özdoğan is a research assistant in the Faculty of Education at TED University in Ankara. She is also a PhD candidate in the Faculty of Education, the Department of Elementary Education at Middle East Technical University. Her research interests focus on teaching and learning of mathematics.

S. Asli Özgün-Koca is a professor of mathematics education in the College of Education at Wayne State University, Detroit, Michigan. She teaches secondary methods course for preservice secondary mathematics teachers, mathematics courses for the undergraduate students, and graduate courses in mathematics education. Her research interests are in the use of technology to enhance middle school and high school mathematics education and mathematics teacher education. She served as the Association of Mathematics Teachers Educators' Technology and Mathematics Teacher Education Committee chair between 2014 and 2016. Currently, she is serving as the president of Michigan Association of Mathematics Teachers Educators (MI-AMTE).

Lisa L. Poling is an Associate Professor in the Department of Curriculum & Instruction at Appalachian State University. Currently, she teaches junior and senior level undergraduate courses in mathematic pedagogy and content and is a member of the graduate faculty. Her research interests focus on teachers' sense of responsibility in educating all children in the science of mathematics and utilizing mathematics to critically assess social issues.

Michelle Stephan is an Associate Professor of Mathematics Education. She has a joint appointment in the Department of Middle/Secondary Mathematics and the Mathematics Department at UNC Charlotte. She taught middle school mathematics in Florida for seven years before joining the College of Education. She teaches middle and secondary mathematics methods classes and provides extensive professional development to teachers in Cabarrus and Kannapolis County. Her research interests include design research methodology, mathematics for the 21st century, lesson imaging for inquiry STEM, and supporting students with special needs in mathematics.

James A. Telese is an Associate Dean for Assessment and Accreditation and Professor of Mathematics Education in the Department of Teaching and Learning, at the University of Texas Rio Grande Valley. He earned a PhD in Curriculum and Instruction with an emphasis in mathematics education at Texas A&M University, College Station. Dr. Telese has experience in program evaluation. He is a member of TODOS-Mathematics for All, the National Council of Teachers of Mathematics, the Association of Mathematics Teacher Educators, the American Education Research Association, and the Research Council on Mathematics Learning. His research interests include national and international mathematics achievement, teacher professional development and self-efficacy, preservice elementary teacher preparation, proportional reasoning, and assessment. Dr. Telese is a former middle school mathematics and science teacher, and a former high school mathematics teacher.

John A. Williams III is an Assistant Professor of Multicultural Education at Texas A&M University at College Station. His research focuses on developing and replicating best practices, policies, and personnel to dismantle inequitable discipline outcomes for African American students in K–12 school environments. Additionally, his research investigates how to prepare and support culturally inclusive teachers through the adaptation of multiculturalist frameworks to instruct in urban. He received his Bachelor's degree in Sociology and Master's degree in Education Policy Studies at the University of Illinois at Urbana-Champaign, and his Doctorate in Curriculum and Instruction at the University of North Carolina at Charlotte.

Dawn Marie Woods is an Assistant Professor of Elementary Mathematics Education in the School of Education and Human Services at Oakland University in Rochester, Michigan. She teaches mathematics methods courses in the elementary education program. As a lifelong educator with teaching experience in PK–6th grade classrooms, she brings a teacher's perspective and curiosity about students' mathematical thinking to her research. This unique perspective allows her the opportunity to investigate the following big idea: How do math language routines provide teachers with ways to foster student participation and advance justice while building math language, practices, skills, and concepts.

Laura Young is a Research Scientist in Temple University's College of Education and Human Development. Her research broadly explores the development of mathematics skills in both formal and informal contexts, how these skills change developmentally, and whether they can predict later mathematics achievement. She is particularly interested in the use of interventions, like worked examples, to facilitate and improve mathematics knowledge.

Deborah Zopf is a mathematics instructor in the School of Science, Technology, Engineering and Mathematics at Henry Ford College (HFC). She teaches classes from developmental mathematics courses through calculus, specializing in the mathematics for elementary teachers courses. She collaborates with teacher educators at several universities to align HFC mathematics. She has worked with educators at Wayne State University on several initiatives.

Printed and bound by CPI Group (UK) Ltd, Croydon, CR0 4YY

15/09/2026

14962336-0003